Web Cartography

Map Design for Interactive
and Mobile Devices

Web Cartography

Map Design for Interactive and Mobile Devices

Ian Muehlenhaus

CRC Press
Taylor & Francis Group
Boca Raton   London   New York

CRC Press is an imprint of the
Taylor & Francis Group, an **informa** business

CRC Press
Taylor & Francis Group
6000 Broken Sound Parkway NW, Suite 300
Boca Raton, FL 33487-2742

Printed on acid-free paper
Version Date: 20130712

International Standard Book Number-13: 978-1-4398-7622-0 (Hardback)

Library of Congress Cataloging-in-Publication Data

Muehlenhaus, Ian, 1976-
     Web cartography : map design for interactive and mobile devices / Ian Muehlenhaus.
       pages cm.
     Includes bibliographical references and index.
     ISBN 978-1-4398-7622-0 (hardback)
     1. Cartography--Computer programs. 2. Cartography--Computer network resources.
3. Mobile geographic information systems. 4. Geodatabases. 5. Digital media. I. Title.

GA102.4.E4M82 2013
526.0285′4678--dc23   2013024769

Visit the Taylor & Francis Web site at
http://www.taylorandfrancis.com

and the CRC Press Web site at
http://www.crcpress.com

To Birgit, Svenja, Antja, and Mette.

Ich hab euch Lieb.

Contents

Preface

The idea for this book came to me three years ago when I was teaching an interactive and Web cartography course for the first time. Many of my students had taken some introductory GIS (geographic information system) courses but had absolutely no coding experience, nor any knowledge of HTML, and often, only rudimentary math skills. On the other hand, a certain subset of the class had programming and math skills but absolutely no knowledge of cartography or design. Long story short teaching interactive and Web cartography was a real challenge.

It remains so today. One of the most confounding issues I keep confronting is that there are few comprehensive texts dealing specifically with Web map design. Those that do exist tend to either be extremely technical or edited volumes that are often too academic for nonexperts. So, instead of using a single text, I typically hunt down journal articles and academic book chapters on a variety of pertinent topics.

Thus, it occurred to me that what I needed was an approachable, comprehensive, and nontechnical text about map design for the Web—not a book about scripting, application programming interfaces, or about designing exploratory tools. Rather, I wanted a book that talked about map communication best practices, a book based on spatial data visualization and graphic design theory. Theoretically, such a book would be approachable enough for desktop GIS users, print cartographers, and university students who have little-to-no experience in Web programming. However, I also felt that the ideal book would be written so that anyone who wants to design effective Web maps could learn about the core concepts of cartography without needing to refer to an additional, potentially more esoteric, source. Essentially, I needed a book that was written by a cartographer about Web map design and written in a language that anyone with even a slight interest in Web mapping could easily understand.

I could not find such a book. So, I decided to write my own.

My goal with this book is to offer a solid cartographic launching pad from which students, practitioners, and innovators can begin to design aesthetically pleasing and intuitive Web maps. With my backgrounds in cartography and map design, I was able to collate and synthesize current Web-mapping norms into this text. My training has also allowed me to critically assess Web mapping within the broader history and science of mapmaking.

So, thank you for picking up this book. I truly hope you enjoy it and that it helps you design more powerful and effective Web maps.

Ian Muehlenhaus
La Crosse, Wisconsin, USA

Acknowledgments

I cannot thank Birgit Muehlenhaus enough for her patience and encouragement, as well as her incredible editing capabilities. (She literally saved you from reading about 100 pages of superfluous text.) How much time she set aside to help me with this book is unfathomable—time that she did not have. On top of this, she has kept our household from falling apart as I spent numerous evenings and weekends locked in the basement with my computer. Thank you, Birgit. This book would not have been finished without you. You are a coauthor in all but title.

I must thank Svenja Muehlenhaus for interrupting my writing binges to listen to the goofy guy and goofy girl sing "Bird'n'Roll" and "John McEnroe." You may not remember this period of your life once you grow older, but just know that I will always cherish my time listening to Dionysos with you throughout the writing of this book. "What's up, le Monde?!"

I would like to thank my mother for always encouraging me to try new challenges. I am where I am in this world because of you.

Thank you, Irma Britton for being so patient and not letting me give up on the project. I really appreciate this and am forever grateful. You are the best editor ever.

I would like to thank my former colleagues in the Department of Geography and Mapping Sciences at the University of Wisconsin–River Falls. It was there that the idea for this book first came to fruition. I miss the daily banter. John, thanks for telling me I should write this book.

I owe a debt of gratitude to my current colleagues in the Department of Geography and Earth Science at the University of Wisconsin–La Crosse. I apologize that my door has been shut so much over the past semester as I wrapped this up. It opens again tomorrow. I am very fortunate to be working with such great people and at such a great university.

I would like to thank all of my students from semesters past and present. I wrote this book because of, and for, you. If it weren't for you, I wouldn't be in academia. Of particular significance while writing this book were Mary Windsor, Derrick Sailer, Hannah Moseson, and everyone from my fall 2012 cartography course. You have all made teaching a real joy since arriving at La Crosse.

I would not be where I am today if it were not for those who taught, inspired, and helped me. Matti Kaups and Roger Miller … I miss you both. Thanks to Gordon Levine for encouraging me to return to academia.

Scott Freundschuh is responsible for getting me excited about map design as an undergraduate student. He was instrumental in providing me feedback during my dissertation, and he continues to help me out careerwise on a regular basis. Scott, I truly appreciate all that you have done for me.

Steven Rosenstone will probably never read this book (he is a political scientist, after all), but after my dissertation defense, he gave me a bit of sage advice that I will never forget. It is because of this advice that this book was written. Thank you, Steven.

Last but not least, I would like to thank my adviser, Robert McMaster. It was an honor being your graduate advisee. I hope I have done you proud. (Oh, and I promise I will have my students still use your, Terry's, and Fritz's textbook, too.)

To all of those I missed, and I am sure there are many, please forgive me. Thank you, too.

About the Author

Ian Muehlenhaus fell in love with maps growing up in Duluth, Minnesota. During the long, dark winters of his childhood he perused atlases and daydreamed about distant, warmer lands. He went on to earn his M.Sc. in geography at The Pennsylvania State University in 2002 and his Ph.D. in geography at the University of Minnesota in 2010. Today, he is fortunate enough to study maps for a living as an assistant professor at the University of Wisconsin – La Crosse.

Ian's research on maps has been published in a variety of journals, including *The Cartographic Journal, Cartography and Geographical Information Science* (CaGIS), *Cartographica*, and *Cartographic Perspectives*. He is the coordinator of the annual CaGIS Map Competition and a former co-chair of the Student Dynamic Map Competition for the North American Cartographic Information Society (NACIS). Ian has also acted as the chair of the Cartography Specialty Group of the Association of American Geographers and is currently an editorial board member of *Cartographic Perspectives* and a map reviewer for the *Journal of Maps*. He has worked as a consultant for the National Geographic Society and National Endowment for the Arts, and been invited to lecture on effective map design at the NASA Goddard Space Flight Center.

Ian's map interests are myriad, although all tend to come back to map aesthetics and purposeful design. Beyond Web mapping, the focus of Ian's research is on systematically designing maps for more effective information recall, likability, and persuasiveness. More recently Ian has become interested in the scholarship of teaching and learning (SoTL). *Web Cartography* was an attempt to write a book that was extremely approachable to a broad audience.

During the academic year Ian resides in La Crosse, Wisconsin, with his wife (fellow cartographer Birgit Muehlenhaus), two daughters, and their Wheaten terrier. In the summer, he and his family are often found in Germany, France, and Hungary visiting family and friends. More information can be found about Ian at www.ian.muehlenhaus.com or on Twitter @iMuehlenhaus.

Ian loves spreading the word about maps and map design! If you would like to have Ian come speak in your neck of the woods or do a Webinar, please don't hesitate to contact him.

1

Introduction

This book is for people who are interested in effective map visualization and communication on the Web. It does not matter if you are a professional cartographer, geographic information system (GIS) expert, computer scientist, or novice mapmaker. The only prerequisite is that you are interested in improving or honing your skills at Web map design.

I wrote this book with two objectives. The first is to synthesize and present a broad overview of which map design processes, issues, and techniques have changed from print cartography to Web cartography and which have stayed the same. That is, this book is an attempt to separate and critically examine what is different between designing a print map and a Web map. Which processes and techniques about map design have changed with Web maps? This question is continually addressed throughout the forthcoming chapters.

Admittedly, however, not everything has changed about map design along with map medium. In fact, many design concepts have proven themselves even more robust when it comes to effective Web map design. This book also reviews numerous established cartographic techniques and concepts that remain in force with Web mapping. Many of these cartographic standards and rules are simply unknown by modern-day Web cartographers. These days, it is almost a sport for trained cartographers to sit around the proverbial water cooler and comment on the work of "neogeographers" for ignoring cartographic truisms such as projection, rules of color, and symbolization. Yet, ask many of these same cartographers to design an effective Web map, and you will often disquiet them because they do not know how.

It is my belief that ineffectively designed online maps exist not because the tools for mapmaking have become ubiquitous and easy to use. Rather, they exist because there are few centralized resources available that focus on how to design aesthetically pleasing, user-friendly Web maps.

Thus, the second objective of this book is to elucidate knowledge about designing aesthetically pleasing and effective Web maps. This goal is achieved by blending relevant concepts from print cartography with contemporary information visualization and Web-mapping knowledge. This book presents a series of Web map design guidelines and suggestions based on the literature and established cartographic standards. Plus, I mix in empirical evidence based on my own experience as a map critic, mapmaker, and instructor of cartography.

Qualifier: This Is Not a Book about Coding (That Is, No JavaScript Required)

It would be unfair for me not to mention what this book *will not* cover in the forthcoming chapters. This book will not teach you the tools and scripts necessary to create your own Web maps (although Chapter 12 offers an overview and discussion concerning some of the tools available to you). This book also will not try to promote one method of Web mapping or software package over another. These topics are technical issues that many other books, online tutorials, and Web resources have and will continue to illuminate better than this book ever could. Finally, this book will not discuss exploratory geovisualization. There are already many books and journals that explicitly deal with this topic. Many Web mappers do not wish to design exploratory geovisualizations. They are simply trying to design effective and useful maps that present information clearly or persuasively. I wrote this book to help people do just that.

The Intended Audience

It is my hope that this book will be of practical use to three broad audiences. First, this book should prove useful to students of cartography who already have training in desktop GIS or print cartography and are now moving into the realm of Web cartography. I recommend working through this book as you learn the technologies in your course work. Second, this book should prove useful to professional cartographers who were trained in print cartography and are now designing for Web cartography. This book is purposefully written in nontechnical, layperson English so that professional and armchair cartographers alike can work their way through it on their own time. Third, this book is targeted at those already making Web maps who have no training in cartography or GIS. This book will offer insights for designing more effective Web maps based on hundreds of years of (still-relevant) cartographic research and the findings of contemporary visualization scientists.

Regardless of your background as a reader, the rest of this chapter is going to get us all caught up on the state of Web cartography. By the end of this chapter, we will all be on the same page. First, a brief history of modern map-making, particularly where and how Web mapping fits into this timeline, is reviewed. This segues into an intense discussion about the importance of mapping with a purpose—a central theme that is referred to in every chapter of this book. The chapter then concludes with a list of key takeaways and additional resources if you desire to read more about any of the topics discussed herein.

Contemporary Mapmaking: A Quarter Century of Rapid Evolution

The art and science of mapmaking continue to change so rapidly that it is difficult to keep up. Merely 25 years ago, mapmakers were focused on how electronic cartography was revolutionizing the industry—no more dark-rooms. With the rise of the Internet in the mid-1990s, cartographic research shifted away from the paper medium to Internet mapping and exploratory geovisualization. In recent years, the rapid adoption of handheld media devices such as smartphones, tablet computers, and e-book readers are allowing mapmakers to design maps that are not only interactive but also as mobile and tangible as paper maps once were.

As cartographic knowledge evolves at a breakneck pace, traditional guidance on what constitutes good map design is starting to show its limitations. Much of this is due to the shift in mapping media from paper to the Web browser and, more recently, to *mapps*—map applications made for mobile devices (Muehlenhaus, 2012). Although some concepts of cartographic design, such as visual variable selection, have remained largely intact with minor addendums (e.g., animation and focus), other conceptual remnants of print cartography have become woefully irrelevant. Traditional rules regarding minimum line thickness and typographic size, in particular, are at best immaterial and at worst detrimental to good design on digital devices.

The primary goal of this section is to introduce you to the history of Web and mobile mapping within the broader context of cartographic history as a whole. The forthcoming pages demonstrate just how different Web cartography is from paper mapping, but they will also impress on you the fact that, for all of the differences, many of the core tenets of cartography remain intact. After providing a history of Web cartography, we delve into different philosophical arguments about what constitutes a map in the modern multimedia context. By the end of this chapter, you will have a thorough and complex understanding of the impact that interactive technologies and the Internet have had on mapmaking.

Web Cartography: A Brief History

The history of cartography is long enough to fill tomes of writing. (This is meant literally, as the history of cartography project that is currently under way already comprises thousands of pages.) To comprehensively analyze how mapping has evolved from primitive reference maps to loading an app on our smartphones is beyond the scope of this chapter. However, a review of the history of modern thematic cartography is pertinent to fuel our understanding of what makes the rise of Web mapping over the past 20 years so revolutionary and distinct from its paper predecessors. The last time cartography went through such a sea change was with the advent of thematic mapping. Indeed, although taken for granted by most cartography students these days as an integral part of mapmaking, thematic mapping is a relatively new enterprise in the history of cartography, beginning in earnest only in the early 19th century. The arrival of thematic mapping in the world of cartography had nearly as dramatic an impact on mapmaking as mash-ups and multimedia maps are having today. So, this is where we begin our modern history of cartography—with the last major paradigm shift.

Thematic Cartography: The Precursor to Multimedia Cartography

Leading up to the 19th century, maps were typically designed and used for demarcating property ownership, state territory, and urban layouts (Pickles, 2004). Maps were used for referencing things. Although thematic cartography as a practice did not yet exist, there were some important antecedents. As far back as the 1600s, scientists had been creating thematic charts of tides, winds, and elevation. Some argue these were the first thematic maps. Regardless, they were missing some key data that we typically associate with thematic maps today: human demographic data.

Thematic cartography as many of us know it today could not develop until data were systematically collected about human populations. Before the advent of modern spying technologies, however, tracking humans was a difficult process. (Now, of course, people willingly allow themselves to be tracked by carrying around mobile phones.) Thus, it was not until the rise of the modern nation-state, with its mandatory censuses, economic trade statistics, and bureaucratic auditing agencies, that there was ample data of interest to map. It was then that engineers, designers, and mapmakers could begin visualizing a variety of nonmeteorological statistics in space, that is, modern thematic mapping.

The French were early experts at thematic cartography. In France, map designers began experimenting with different methods of data visualization, including dot maps, flow maps, proportional symbols, and choropleth maps. They used the vast amount of data made available to them to begin mapping economic trade data, diseases, population characteristics, and perhaps most famously, casualties of war over time (Figure 1.1). States and businesses soon realized how useful these cartographic visualizations were to understand the nature of their populations as compared to tables full of numbers. These maps could also be used to investigate spatial patterns occurring within a population at large. Probably the most famous example of this in English-speaking countries is John Snow's cholera map, used to help investigators in London link the outbreak of cholera to bad well water (Figure 1.2).

From the beginning of the 1800s through the early 1930s, thematic cartography was largely undertaken by architects, engineers, and designers—not those specifically trained in map design. Cartography is a relatively new term that was not even coined until well after thematic maps were becoming more common. Academic cartography—or the systematic study of visualizing data spatially—is relatively new (Wood, 2003). Often, the designers of early thematic maps taught others their techniques, but these transfers of knowledge were typically not done in a formal academic setting; they were instead more like apprenticeships. Mapmakers handed their knowledge down to others who were interested in learning how to make maps. Leading into World War II, it soon became apparent that maps were not only useful tools of objective visualization but also were rhetorical tools that could be used to spread propaganda. Immediately after World War II, more universities began offering courses in graphic and cartographic design (McMaster & McMaster, 2002).

Once cartography became engrained in academic institutions, it gained scientific legitimacy. With legitimacy, graduate programs in cartography began to flourish. Thus, cartographic knowledge was institutionalized and passed on in a systematic fashion—via academic degrees. Map creation was not a simple process. It required a mix of mathematical knowledge to scale and project one's maps appropriately, as well as the ability to draw and write impeccably. Finally, map production often meant many hours in a darkroom, essentially mastering photograph manipulation and reproduction. Cartography truly was a mix of art and science; some opine that it was more a craft than academic endeavor (Case, 2007).

The Rise of Multimedia Mapmaking

Today's cartographic situation may parallel that of the 1800s; we have been making one type of map—paper maps—and now weekly there seem to be new technologies and media that are revolutionizing what maps are and can do. Universities still have cartography courses, but most of these courses deal with paper methods of map design. Likewise, principles of cartographic design are often given short shrift within GIS courses dealing primarily with spatial analysis. Academic cartography simply could not keep pace with the rapid changes taking place in technology and data collection since the late 1980s. In essence, it seems that the craft of online mapmaking has developed without too much input from the discipline of cartography itself.

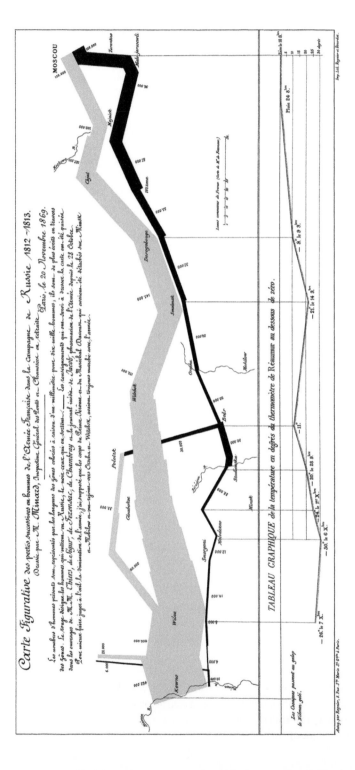

FIGURE 1.1

Carte figurative des pertes successives en hommes de l'Armée Française dans la campagne de Russie 1812–1813, by Joseph Minard, 1869.

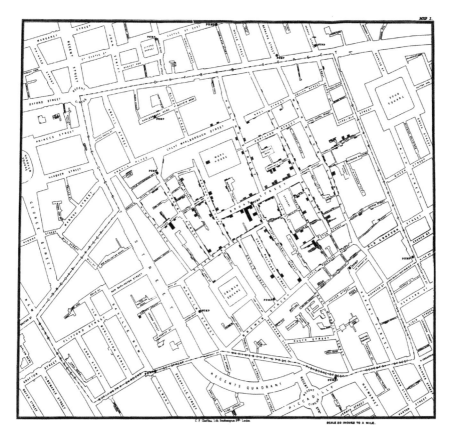

FIGURE 1.2
Dot map by John Snow created after the London cholera epidemic of 1854.

Today, a majority of online and mobile maps are created by computer scientists, Web designers, or self-taught coders. These mapmakers are experimenting with different interactive mapping APIs (application programming interfaces), such as Google and Bing Maps, and visualization techniques much as engineers and architects dabbled with thematic cartography in 19th-century Europe. Online mash-up maps (i.e., those push-pin or upside-down teardrop Google maps) are a prime example of this tendency.

This will remain the case unless academic programs can synthesize the massive amount of new knowledge dealing with online and interactive maps and contribute back to those actually designing maps (something this textbook is a first attempt at doing). But, how did online mapmaking so quickly spin off and away from the discipline of cartography? The change from paper to digital media was quick and systematic and is ongoing. It began with the rise of hypermedia.

What Happened to Paper?

Paper is a static medium. One cannot really interact with the data presented on a piece of paper. A paper map must be designed with the understanding that whomever uses it will not be able to simply turn off layers to see what he or she wants. Paper maps must be presented with care so that relevant pieces of data are not lost in the sea of information that is a map. One reason academic cartography has flourished is because learning how to flatten data from spherical Earth onto a piece of paper, and presenting said data in a format that is palatable for map readers, takes special training. Print cartography is part art and part science. For centuries, paper was the undisputed medium of choice for mapmaking. It had many benefits over other media. First, it was more portable than most things; hauling around maps carved into stone quickly grew burdensome. Second, with the advent of the printing press, paper maps could be mass produced quickly and distributed widely. Finally, over time paper became one of the cheapest media to produce. That is, the cheapest until the advent of hypermedia.

Early Hypermedia

What is hypermedia? Before answering this question, we need to understand its origins. Hypermedia was preceded by the rise of personal computing and online bulletin board services that used hypertext. Hypertext is the linking together of digital text documents via common keywords or phrases that allow a reader to go through a document in a manner that is not unidirectional. In essence, it is whenever a person can click on a word or phrase and be taken to additional information related to that word or phrase. Hypertext is taken for granted today, but it was an incredible innovation that revolutionized how we consume written materials. Instead of reading documents straight through, as was often dictated by the paper medium, digital interfaces allowed links tying certain topics and themes together to be embedded throughout a document. These links gave readers some control over how information was fetched and obtained. The early Internet, comprised almost entirely of text, was built on hypertext—indeed, HTML stands for Hypertext Markup Language.

   Hypermedia, however, goes a step beyond hypertext. As digital and computing technologies increased in power, there was no longer a need to limit linking to text. Hypermedia is the extension of hypertext through multimedia, including maps, images, sounds, animations, and videos (Cartwright & Peterson, 2007; Cartwright, 2007; Peterson, 2007). It allows for a truly multimedia experience that combines all types of data presentation in a manner with which a user can interact. During the 1990s, hypermaps became increasingly prevalent. Hypermaps were maps that contained links to other types of media (e.g., national anthems, flags, pictures, graphs, and textual information) (Cartwright & Peterson, 2007; Cartwright, 2007). These maps were digital, with no paper equivalents. They allowed the user to interact and explore available data, but the maps themselves were largely limited in functionality.

One reason for this limitation was that before the widespread diffusion of the Internet, early hypermaps were often produced and distributed on discrete media (i.e., compact disks) (Cartwright, 2007). These disks would contain a variety of multimedia documents and data that would be presented in a graphical user interface of some sort. For example, beginning in 1996 the *CIA World Fact Book* was distributed on CD-ROM. You could load it on your personal computer (PC), interact with it to find information regarding various parts of the world, load pictures of national flags, and find maps. Today, however, you will not find this resource on a disk, and even if you do, you would be a fool to pay for it as by tomorrow, the information about many of the countries will be obsolete and already updated online. The Internet ushered in the ability for people to update datasets and graphics rapidly and for the distribution of these updates to be received near instantaneously. The diffusion of the Internet into homes and offices changed the medium of cartography forever. It shifted from discrete (e.g., paper, CD-ROM) to ephemeral, cloud maps.

Why Hypermedia Evolved into Web Mapping

The way it has come to dominate society, economics, politics, and even our private lives, it is amazing that the Internet has only existed in its present, commercialized state for 20 years. As recently as the mid-1990s, map design courses were obsessed with how easy it was becoming to make paper maps due to the melding together of drawing programs, GIS, and personal computers. Within 10 years of this electronic revolution, cartography had to weather and adapt to the rise of Internet and multimedia technologies. Two hundred years of acquired knowledge on making maps for the paper medium, a medium that seemed destined to reign supreme for decades to come given the ability of computers to produce high-quality paper maps, suddenly became largely irrelevant.

The Internet spelled the end of discrete, machine-based multimedia. As more and more pieces of technology—not only PCs but also televisions, mobile phones, and even digital picture frames—became attached to the Internet, discrete media began showing limitations. Whereas you could store and distribute a map on a disk, eventually that data would become out of date. It was cheaper and easier to have someone install a program that loaded and mapped the newest data available straight from the Internet. Over time, as Web browsers increased in interactive capability—often via plug-ins such as Macromedia (now Adobe) Flash—discrete media fell by the wayside completely. Suddenly, people just surfed to their favorite Web sites to get information, interact with multimedia, or use maps.

Literally within 10 years of the Internet going commercial on January 1, 1994, the wired world of 56K modems had transformed into a high-speed, wireless, and ubiquitous component of modern society. Desktop computers and, in the antiquarian parlance of the early 1990s, workstations were replaced by more portable laptops. Files and programs were increasingly stored on company servers for workplace-wide access. Eventually, programs and files moved to larger cloud servers, allowing for easy access to data and information regardless of where someone was. As computing increasingly becomes something done online, so does mapping.

"Web," "multimedia," "online," "interactive," "Internet," and "hypermapping" all became buzzwords among cartographers by early 2000. Academic debates arose over the idiosyncrasies between these different map styles and the proper definition of each. With online and digital mapping, new types of maps began to flourish, using visual and other variables (e.g., animation and sound) heretofore impossible to use on paper maps and older PCs. New theories on cartographic representation and techniques were developed. And then, within five years, technology pushed the mapmaking envelope again, redefining what we mean by interactive map design.

How Google Maps Revolutionized (or Was It Euthanized?) Mapmaking

In 2005, Google unleashed Google Maps on the world, and cartography would never be the same. Google Maps and a variety of other companies' similar offshoots (including Bing Maps, Yahoo! Maps, and more) allowed for people with absolutely no cartographic training to easily create their own maps. Not only could they find directions between places, but with access to the API and Google's acquisition of KML (Keyhole Markup Language), people could even create their own data layers to be placed over Google's base maps. Further developments, such as Google Earth and the ability to upload and sync one's photographs of a place using online maps, meant that interactive mapping was becoming increasingly bidirectional. Not only could a person interact with a map but also the person could add to it and change it. This development paralleled the growth of Web 2.0 technologies. Web 2.0 represents the shift in Web design from mere user interaction with Web sites to user interaction with and manipulation of Web sites. Essentially, the development of map mash-ups—copyrighted base maps with other people's data overlaid—have allowed Web users to become content creators.

Although liberating, do-it-yourself mapmaking has had some serious side effects on quality—or what we might refer to as the Googlization of maps. First, due to their ubiquity, people have come to expect that all online maps *should* look like Google Maps. Although Google Maps introduced many innovative map interface features—such as a pan-and-zoom toolbar and map tiles for quick reloading of data after scale change—the interface and map design Google uses is not infallible. In fact, it suffers from a variety of shortcomings that are referred to throughout the rest of this textbook.

A second problem with our obsession with the Google Maps style—it inherently promotes inaccurate representations. Just because Google Maps-style mash-ups are easy to make and ubiquitous does not mean they are good. For example, nearly all Google Map mash-ups, at least those not hacked by programmers, use a Web Mercator projection. The reason for using one of the least-useful projections ever invented—outside of ocean navigation, perhaps—is simple. It is a rectangular projection. Google wants the map to take up the entire browser. As an Internet service provider, Google is not troubled by the fact that this projection distorts the area of landmasses to the extent that thematic visualizations should not be represented on it. People will use it because it is easy, and it is also now synonymous with online mapping. On the plus side, the Web Mercator projection shows minimal distortion at large scales. This is great for Google when it is providing driving directions around a city, but not so useful for designing a thematic mash-up of East Asia.

A third limitation with Google Maps mash-ups is that the base maps are created and generalized by Google. Although you can change the type of background map (e.g., aerial photography, base map, terrain, and more), you cannot easily weed out roads or small-scale water bodies or remove other features that are not relevant to what is being mapped. From another standpoint, one also cannot decide what to include on the base map. Often, Google generalizes out features that might be relevant to the theme being mapped.

This leads to the final implication of the Google Maps hegemony: the power to delete things from existence and distort reality. As is the case with all maps that the public accepts as official and legitimate, if something does not exist on a Google Map, it will cease to exist in people's mental maps of places unfamiliar to them. Even more worrisome perhaps is that if something does exist on a Google Map, people perceive the map as having more legitimacy than the facts on the ground. Numerous studies show that people will hold to their beliefs (e.g., the accuracy of Google Maps) even in the face of overwhelming evidence to the contrary (e.g., Google Maps is not right). This may sound ludicrous, but a Google Maps misrepresentation actually caused an international invasion. In 2010, Guatemala invaded Costa Rica after a commander mistakenly thought that Costa Rica was occupying an island that rightfully belonged to Guatemala (Swaine, 2010). Google Maps had incorrectly represented the border. For better or worse, Google Maps has increasingly become the reference map of the world.

The Future of Web Cartography May Be Browser-less

Academic cartography has finally begun to catch up with developments in Web mapping. More and more articles have been coming out on designing cognitively salient and accurate mash-ups and animated maps. Best design practices for certain mapping situations are sprinkled throughout different journals, edited volumes, and online tutorials (to find many of these, please be sure to peruse the Further Reading sections at the end of each chapter). One of the goals of this book is to synthesize this information within the forthcoming chapters. Yet, just as academic cartography seems to have caught up with developments, another extraordinary development began occurring. Web mapping is transitioning again, away from the PC and onto mobile, handheld devices.

Although this book refers to Web mapping throughout, I encourage the reader to think of different ways in which today's maps can be designed for different scenarios, including those when an Internet connection is not available. The rapid transition from paper to discrete media, discrete media to hypermedia, hypermedia to Web, and now Web to simply mobile mapmaking begs a question: What is a map? Has the definition of what a map is changed along with the technology that serves as a map's foundation? Are all of these maps discussed so far even the same thing? The next section briefly explores a key concept that ties all types of maps, both static and dynamic, both print and Web, both browser based and app based, together.

The Goals of Map Communication Remain the Same

All maps are a form of geocommunication. They are all designed to communicate something about our spatial environment to a map reader or user. Most maps are created to help illustrate information of some importance to a particular audience. Others are used to make rational or rhetorical arguments about the state of our environment. Medium does not matter; interactive technologies have changed many things, but they have not changed the fact that most maps are designed to communicate and reveal information, knowledge, or an agenda to audiences.

Geocommunication is at the core of defining what a map is because it exemplifies what a map does. However, *effective geocommunication* depends on something very specific. Maps, or more specifically *excellent maps*, are designed with a communicative purpose. A map that merely represents data is no more useful than an encyclopedia. The data might be articulately presented, but if no arguments are made, no signals or directionality are provided to the reader about what the data imply or might be useful for, a reader will have a hard time getting anything worthwhile out of a map. All design decisions, every single one, should be made with monastic obsession toward achieving your communicative goals. All maps, regardless of type, need to be designed to communicate a message, information, knowledge, or argument in as effective a manner as possible. This is a central theme throughout this book.

The importance of mapping with a purpose is of particular significance in a book about Web mapping for two reasons. First, we are living in a data-rich society, which means we have to decide which data are, or better still, which information is, worth mapping. Digital data are everywhere and are easy to create, get, find, steal, and distribute. We can collect quantitative data about something as trivial as how fast we eat via an Android fork (Gilbert, 2013). We can now quickly access stolen, previously inaccessible, data on sites like Wikileaks. We can share and distribute data around the world with torrents, dropboxes, and a variety of other means. At the same time, Web mapping technologies have become ubiquitous, making the mapping of any spatial data we want particularly easy. Quite simply, though, many data do not deserve to be mapped, just as they do not always deserve to be graphed; mapping them serves no purpose.

This is not to say that mapping data just for the fun of it will not result in wonderful-looking visualizations. Indeed, many data visualizations are stunningly beautiful. However, without a purpose or goal, maps become either visual encyclopedias or abstract art. They tend to be anecdotal and do not effectively communicate much of anything. Similar to viewing random Twitter feeds, you may discover information you did not know, but this unstructured information will not add to your knowledge base in any useful way. Well-designed, data-rich maps that provide no argument or do not have a clear communicative goal are like cute cat photographs on Facebook: nice to look at but worthless.

The second reason mapping with a purpose lies at the heart of this book—and is repeated ad nauseam throughout the upcoming chapters—is that Web mapping makes it easier than ever for map designers to be distracted from their communicative goals. Increasingly, map communication has become a two-way process. No longer are we designing information for someone simply to read and interpret; map users expect to be able to interact with our map to receive the information in myriad ways. This can be a good thing. Interactivity potentially allows the message of the map to reach a broader audience. Map designers can build redundant interactive mechanisms into their maps (e.g., multiple methods of zooming or changing data layers) to help people digest the information they are receiving in individually more palatable ways. However, such interactivity is only effective if and when it is designed and included to achieve the mapmaker's communicative goals. Well-designed interactive tools in Web maps help reinforce the message that is being communicated. Interactive features included merely because they are considered "cool" will distract and take away from the message a mapmaker is trying to convey. In sum, poorly designed Web maps are often the most interactive. They provide a ton of superfluous interactivity without adhering to the communicative goals of the map—without recognizing the map's purpose. They succumb to style over substance. To borrow a phrase from Tufte (1983): Excessive interactivity is to Web map design what "chart junk" is to print.

Fortunately, there is a straightforward process one can go through to design effective Web maps without losing sight of map purpose. It involves determining a future map's raison d'être before beginning a project and consistently returning to this throughout the design process. This is as true for Web and mobile map design as it is for paper maps. However, in the forthcoming sections, I review the process as it pertains to Web map design.

Establishing a Communicative Purpose for Your Map

Maps presenting information do more than merely show things. Just as a news article presents information about what is going on in the world, maps assert visual statements about how the world is (Koch, 2004). The first thing a map designer needs to do is dispel the notion that maps can be objective visualizations of data. They cannot be. They may be designed neutrally, but as soon as you put north at the top or choose a projection, objectivity goes out the window. Maps designed for presentation are always visual arguments that attempt to inform, and indubitably end up reshaping, a person's mental map of the world. Maps help people make spatial sense of the world around them—aspects of the world that are too complex to comprehend via life experience and our five senses alone. Your maps will shape how people see and interact with the world, similar to how movies and television shape—or, some would argue, distort—people's views of society. Therefore, design elements should only be used to enhance a map's message. Always be wary of accidentally burying a Web map's message with unnecessary interactivity and on-screen gadgetry.

For a map to maximize its communication potential, mapmakers need to ask themselves four questions throughout the design process:

- Who is the intended audience for this map?
- Which data will best communicate these things?
- What are the one to five things I want this map to communicate clearly?
- Which design elements will help the map user receive the message most clearly?

When mapmakers forget to ask themselves these questions before and while they design a map, bad things happen.

Who Is the Intended Audience?

The most important question you can ask yourself before beginning to produce any map is: Who will view this? Notice, you are not asking, Who might view this? or Who could potentially view this? You need to have a target audience in mind when you design a map. There is no one-size-fits-all method to any type of communication. Although you will certainly have no control over the diffusion of your map on the Internet or in mobile app stores, you can have control over shaping the map to communicate effectively to a particular target audience. This should be your first design goal.

The intended audience, above all else, will dictate the rhetorical style and design decisions you make underlying your project. Establishing your map's target audience provides a road map for any design decisions you make. Never target your map at "everyone" as it increases the likelihood that your map will not resonate well with anyone.

Map context plays an important role as well. Will your map be embedded in a Web site or appear as a pop-up window in a Web browser? Perhaps you will be designing two versions of the same map—one that is created specifically for a Web site and another for mobile devices such as smartphones and tablets. These are just several reasons to consider the medium and context of your map before you begin producing it.

What Data Need to Be Communicated?

Listing what you hope to accomplish with a map can be one of the best ways to hold yourself accountable throughout the design process. You want to be sure that the key points of your map, and the information that needs to be communicated, will be clear. If you just jump into the design process without thinking about these points, you will likely diverge into fanciful design tangents that look neat but have nothing to do with the message you are trying to communicate.

How Do I Design My Map So Information Is Easily Recalled?

One of the first rules of teaching is that people never retain everything they have learned. A year after taking a 15-week course, a person will likely only recall in any detail three to five topics that were covered. These will likely be the topics that the instructor returned to repeatedly throughout the course, or that were so interesting and influential to the student that the student cannot help but remember them.

Before a semester begins, teachers need to decide which three to five topics they want their students to retain after the semester is complete. For example, I realize that many of my geography students who take my introductory cartography course will never go on to make maps. My goal therefore is to saturate my students' minds with particular truisms that they will not forget. Even if a student does not remember the exact nomenclature used to describe the different projections, my goal is to make sure the student will never forget which projections are inappropriate in particular circumstances. To this end, I continually critique maps shown and designed in class regarding projection use and appropriateness.

As a cartographer, you must do the same thing. You need to ask yourself: Which three to five pieces of information do I want map users to have ingrained in their brain when they are done looking at my map? With reference and location-based service maps, you will often let the user decide what he or she is looking for (e.g., nearby restrooms), and once the user selects that information, you will design the map to highlight it effectively. With thematic data, though, you will often need to design a map that shows a map user what the user needs to know. This is done by highlighting certain data in the visual hierarchy, weeding out extraneous data, and generalizing data.

Before beginning to design any interactive map, it is wise to list what you want the map user to take away from the map. If the end product is to be a user-driven reference map, you should list the types of data a user will be able to call on (e.g., restrooms, restaurants, pubs) and decide how you want to emphasize the data the user wants. Your list, sadly, should be no more than three to five themes, although within these themes you may have nuanced data. For example, you may have a restroom theme that is broken down into categories: pay public restrooms, free public restrooms, private restrooms you can probably sneak into, and bushes providing good coverage in a pinch. These would all comprise one theme. With thematic data, though, it is particularly imperative that you decide what themes you want the map user to see and interact with so that you can make sure the user leaves with the knowledge you are trying to present.

This will not be your only list, however. Generally, there are many tertiary types of data and themes that tie into what you are trying to communicate. For example, a weather map may benefit from elevation data, even though elevation is not what the map user will find most pertinent when looking to see if a storm is coming. You should also make a list of data that would be useful to map users given the purpose of the map. Data and themes from this list should be referred to throughout the design process. Eventually, many of these datasets may be incorporated into your map in well-designed ways so they do not take away from the main message of your map but will be there to assist the map user. When it is time to make decisions about including certain components and data, however, things on the primary list should take precedence over those on this secondary one. (This is reviewed more comprehensively in Chapter 3.)

Why make lists? Although it sounds like advice you would receive from a self-help book, having lists differentiating what things you *must* and *would like* to communicate will guide you through the entire map design process. (You may even consider pinning these lists to your OS desktop via a digital sticky note, as in Figure 1.3, as you will want to refer to them at every point in the design process.) These lists will begin to inform your decision making about a map immediately. Decisions dealing with projection, base map generalization, data selection, symbolization, interactivity, map element design, and many other aspects of the map can be easily made by referring back to these lists.

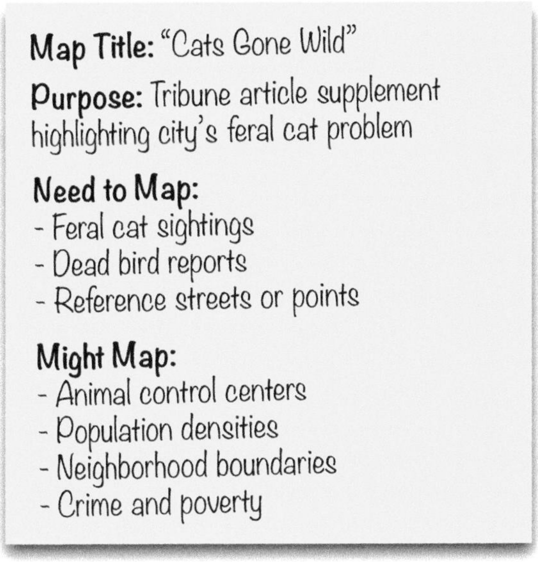

Map Title: "Cats Gone Wild"

Purpose: Tribune article supplement highlighting city's feral cat problem

Need to Map:
- Feral cat sightings
- Dead bird reports
- Reference streets or points

Might Map:
- Animal control centers
- Population densities
- Neighborhood boundaries
- Crime and poverty

FIGURE 1.3

An example of a reminder note. I recommend using these. They can be effective at helping you stick to the communication goals of your map so that you do not start mapping or adding things to your map just because you can.

How Do I Determine Which, and How Much, Data to Include?

It is always good practice to think about which data *need* to be included versus which *could* be included early on in the design process. Data selection all ties back to map purpose. Now that you have a list of the major themes your map must show to communicate effectively, you can begin narrowing down what data you actually need, and how much of it, to make an effective map. Having these lists will help you begin searching for data that will actually make it onto your map, rather than combing for any and all data that "may" be useful for your mapping needs.

Too often, Web mapmakers attempt to put every type of data they can find into their map, even though the map itself is meant to communicate, not be used as an exploratory device. Too many choices, when it comes to map design for presentation, decay communication. Map purpose must guide data selection. Just because you can include data does not mean you should. It very well may detract from your message. Remember that a map user is likely going to take away three to five themes or ideas from your map. By including a variety of menu buttons and datasets that distract from your main message, you minimize the effectiveness of the message you are attempting to communicate.

Designing Your Message

Web mapping and print mapping are different, but several things remain unchanged: They are both forms of geocommunication, and the most effective maps make clear arguments about the information they are presenting. Maps designed with a communicative goal (i.e., purpose) are the most effective. Regardless of the medium, maps providing organized information or knowledge are more useful and easier to make sense of than those just mapping myriad data. Web maps should be interactive, but they should only include interactivity that helps achieve the particular communication goal. They should never be interactive just because a mapmaker is technically skilled enough to make them so.

The Rest of This Book

With this key knowledge now firmly entrenched, the forthcoming chapters provide information and concepts to design effective Web maps. Each chapter reviews the slight changes between print and Web mapping before diving into what has changed significantly. The chapters also review best practices and things to think about when designing Web-map-specific components. This book is concerned with helping you design better Web maps, and as such, chapters are largely broken down by topics that deal with specific map components.

Chapter 2 reviews different types of human-computer interactivity and discusses the opportunities (as well as pitfalls) that such interactivity opens to you as a map designer. Chapter 3 reviews how traditional map elements have mutated into graphical user interfaces. Best practices for designing many different elements are considered. Chapter 4 covers the concept of visual hierarchy as well as different map layout techniques to make the most of your screen real estate in Web browsers and on mobile devices. Chapter 5 discusses color theory as it pertains to Web map design. Much of this may be review for print cartographers, but at the same time, it is hoped that this chapter fosters a more critical analysis of the use of color in Web mapping as contemporary Web maps often break cartographic rules regarding effective color use. Chapter 6 discusses how typographic design and aesthetics have evolved from print maps and reviews common limitations that arise when producing text for screens. Chapter 7 provides a brief review of what visual variables are and when you should use them. Chapter 8 discusses the importance of symbol design for intuitive communication. Chapter 9 reviews many of the most common map representation techniques that are used in contemporary Web maps, as well as several techniques that might be considered for more accurate interpretation. Chapter 10 covers map animation, including additional visual variables you should be aware of and best practices. Chapter 11 addresses the use of sound and haptic feedback in your maps. Finally, Chapter 12 covers the basics of Web mapping technology and also provides an assortment of software, service, and online tools that you might consider using to design your Web maps. Beyond this book, additional information, PowerPoints, Web lectures, video tutorials, and links to external resources will be made available to owners of this book at the accompanying Website: www.ian. muehlenhaus.com/webcartography.

Key Concepts

- This book is about map design for the Web. You *do not* need to know how to make a Web map before reading this book.
- You *will not* be taught how to use technology in this book. The content is concept based.
- Although Web mapping represents a dramatic shift from print mapping, many of the core principles of mapmaking developed over the past centuries remain important.
- Web mapping is a form of multimedia mapping. It is dynamic and interactive (most of the time) and makes use of or connects to many different types of media.
- Map purpose must drive every design decision you make when building a Web map.
- Do not worry about making your map appeal to everyone, just to your intended audience.
- Limit what you are trying to show on your map. What are the one to five things you must communicate? Emphasize these. All other data are expendable if they do not facilitate your communication goals.
- Do not add map elements and interactivity just because everyone else does. Only add interactivity that will better help your intended audience receive the message you are communicating.

Further Reading

Cartwright, W. (2007). Development of multimedia. In W. Cartwright, M. P. Peterson, & G. Gartner (Eds.), *Multimedia cartography* (2nd ed., pp. 11–34). Berlin: Springer.

Peterson, M. P. (2007). Elements of multimedia cartography. In W. Cartwright, M. P. Peterson, & G. Gartner (Eds.), *Multimedia cartography* (2nd ed., pp. 63–73). Berlin: Springer.

Peterson, M. (2008). Maps and the Internet: What a mess it is and how to fix it. *Cartographic Perspectives, 59*, 4–11, 67.

References

Cartwright, W. (2007). Development of multimedia. In W. Cartwright, M. P. Peterson, & G. Gartner (Eds.), *Multimedia cartography* (2nd ed., pp. 11–34). Berlin: Springer.

Cartwright, W., & Peterson, M. P. (2007). Multimedia cartography. In W. Cartwright, M. P. Peterson, & G. Gartner (Eds.), *Multimedia cartography* (2nd ed., pp. 1–10). Berlin: Springer.

Case, N. (2007). Taking apart cartography: our field as a graphic tradition. Paper at the North American Cartographic Information Society Annual Conference, October, St. Louis, MO. http://nat.case.home.mindspring.com/nacis07G.pdf

Gilbert, J. (2013). HAPIfork: Buzzing fork offers ultimate first-world solution to overeating. *Huffington Post.* Retrieved from http://www.huffingtonpost.com/2013/01/08/hapifork-buzzing-fork-solution-overeating_n_2433222.html

Koch, T. (2004). The map as intent: variations on the theme of John Snow. *Cartographica, 39*(4), 14.

McMaster, R. B., & McMaster, S. (2002). A history of twentieth-century American academic cartography. *Cartography and Geographic Information Science, 29*(3), 305–321.

Muehlenhaus, I. (2012). From print to mobile mapps: how to take Adobe Illustrator maps, add pinch-to-zoom, and place them on the Android market. *Cartographic Perspectives, 69*, 59–70.

Peterson, M. P. (2007). Elements of multimedia cartography. In W. Cartwright, M. P. Peterson, & G. Gartner (Eds.), *Multimedia cartography* (2nd ed., pp. 63–73). Berlin: Springer.

Pickles, J. (2004). *A history of spaces: cartographic reason, mapping, and the geo-coded world* (p. xxii, 233 pp.). London: Routledge. Retrieved from http://www.loc.gov/catdir/toc/ecip042/2003008283.html

Swaine, J. (2010). Google Maps error sparks invasion of Costa Rica by Nicaragua. *The Telegraph.* Retrieved February 18, 2013, from http://www.telegraph.co.uk/news/worldnews/centralamericaandthecaribbean/nicaragua/8117902/Google-maps-error-sparks-invasion-of-Costa-Rica-by-Nicaragua.html

Tufte, E. R. (1983). *The visual display of quantitative information* (p. 197). Cheshire, CT: Graphics Press.

Wood, D. (2003). Cartography is dead (thank, God!). *Cartographic Perspectives, 45*, 4–7. Retrieved from http://makingmaps.owu.edu/mm/cartographydead.

2

Human-Map Interactivity

Introduction

The biggest difference between designing maps for print versus the Web is that we no longer design for map *readers* but map *users*. This change in terminology is important. People interact and manipulate Web maps. We no longer need to attempt to design a one-size-fits-all, optimal form of data communication. These days, it is imperative that we design our maps to be interactive and responsive to a map user's needs to facilitate communication that is more effective. Sure, map readers could also manipulate and interact with paper maps. Paper maps can be moved further away or closer to one's eyes, trimmed with scissors, and crumpled to fit in a glove box, but the data on the map stay exactly the same. With most Web maps, users can actually interact with (i.e., manipulate and change) what they are viewing. In sum, interactivity is powerful, and when graphical user interfaces (GUIs) are well designed, human-computer interaction (HCI) can be extremely useful.

Of Mice and Touch Screens

Interactivity with a Web map was a fairly straightforward endeavor a decade ago. Essentially, if you had a computer, you had a mouse. Thus, GUIs facilitating HCIs were almost always designed using the WIMP (windows, icons, menus, pointer) model. In turn, multimedia and Web maps were also designed using the WIMP interface model.

WIMP interfaces (e.g., Microsoft Windows and Mac OS) rose to prominence beginning in the 1980s, and before long, they quickly began replacing many command-line, text-input interfaces (e.g., MS-DOS, UNIX, Basic). WIMP interfaces have proven extremely robust, as they almost always mimic a two-dimensional desktop space to which nearly everyone can

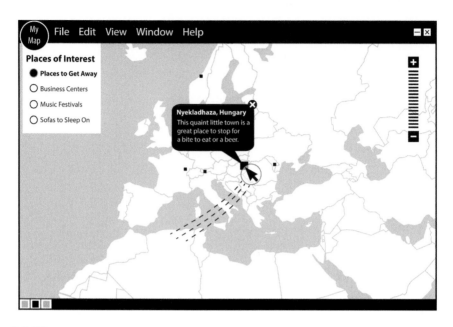

FIGURE 2.1

This is an example of a WIMP interface. The map is displayed in a window area. Icons here are found with the minimize and close buttons in the upper-right corner of the map interface, as well as the close icon on the info window and the radio buttons in the drop-down layers menu. Menus are found on the top part of the map. The pointer here is represented with a standard arrow cursor. It can be moved around the screen and be used to click or hover over icons and menus. In this case, it is hovering over the point symbol for Nyekladhaza, Hungary, at which time the info window opens.

intuitively relate. The four components of WIMP interfaces are so well known today that they are taken for granted (Figure 2.1). *Window* refers to the frame line around a functioning program. (It is often resizable within whichever operating system it is running.) *Icons* are buttons that perform a task. These buttons are typically designed as images or pictograms (e.g., print). *Menus* can be either graphic or text based. They function much the same way as icons; however, they always execute commands, programs, or run tasks. A *pointer* is an always-on, user-controlled, virtual point on the screen (most often visualized with an arrow) that allows someone to interact with the GUI.

WIMP interfaces are still ubiquitous but will soon find themselves in the minority. Even before the arrival of mobile computing, we began to see the rise of post-WIMP interfaces. The early iPod is one example of a post-WIMP user interface. With mobile phones and tablets, as well as touchscreen PC operating systems such as Microsoft Windows 8, pointers are completely disappearing. Google's Android and Apple's iOS, for example, are not WIMP interfaces; they are based on multitouch input, allowing for more

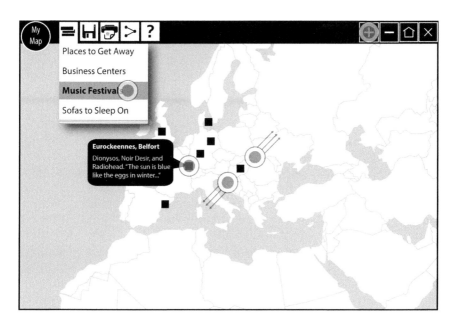

FIGURE 2.2

This is an example of a multitouch post-WIMP interface. There is no ability to hover over objects as touch and drag are the only gestures available. This opens up the opportunity for many other types of interaction, however, including pinch to zoom (over Nyekladhaza, Hungary) and easy map panning. On the other hand, the interface needs to be redesigned to be finger friendly as fingers are not as accurate as pointers. Larger icons have been made to replace the old text-based menus, and place symbols have been made larger as well.

than one point on the screen to be interacted with at a time (see Figure 2.2). The future is post-WIMP, and as a Web cartographer you need to prepare and plan for this. Any map you design today should probably be WIMP compatible but always post-WIMP user friendly.

What does this mean? Essentially, you will want to include some GUIs for pointer devices (such as zoom bars, further discussed later in the chapter), but you should also build post-WIMP capability into your map (e.g., pinch to zoom, finger panning, map rotation, and GUIs that are chubby-finger friendly). That way, on a mobile phone a user will not need to hit the zoom bar perfectly with a thumb to get your map to work. On the flip side, if the same person is opening your map on a friend's desktop personal computer (PC), he or she can still use your map because it will have a zoom bar.

Not everything about a WIMP map interface is antiquarian. Far from it— many of the GUIs designed for computer mice, such as navigation buttons with rollover effects, are still around and will be for years to come. The reason? They have become standard elements found on most Web maps, and they also work with touch-based devices. In many cases, some interactive map elements have come to define maps in general. For example, on the first day of class each semester when I teach my introductory "Maps and Society" class, I ask students to name what elements all effective maps should have. Recently, the first element mentioned by many was a "zoom bar." Zoom bars did not even exist 20 years ago. Now, they are considered one of the most crucial elements of a map—not necessarily Web maps, mind you, but maps in general. I was taken aback. What is ironic is that within 10 years the zoom bar will probably become largely obsolete, as it is simply not necessary with multitouch devices.

The point of this digression is this: Many of the Web-mapping widgets, map elements, and norms from the WIMP interface era are going to be with us for a long time. It will be important to honor established Web-mapping conventions and norms so that your map users do not become confused and frustrated. However, we also need to plan ahead and push the envelope to make full use of post-WIMP features. This means we have to include not only WIMP interface components but also the option of post-WIMP interface components. It is an exciting era of experimentation. Even multitouch finger input is largely passé these days; voice control is already very real on mobile devices, and there is increasing evidence that motion-sensor and eye-tracker user input are going to become far more ubiquitous in the near future. Will your map be able to interpret someone doing the moonwalk as a panning gesture?

New Interactive Map Elements

First things first: traditional map elements and their design are still crucial for effective map communication. The main difference is that most map elements can now be better thought of as GUIs. Thus, the purpose of most map elements may not have changed, but rather their design has become very different. For example, many map elements for Web maps can be designed far more compactly and subsequently present more information once someone has decided to interact with them.

In the following chapters, I outline three key fundamentals to remember when designing interactive map elements. These remain unchanged from print map design. First, find and design map elements that facilitate your communication goals rather than simply using default map elements. Second, not all maps need every map element. In fact, sometimes map elements clutter a map and its message. Be selective. Third, covered in Chapter 4 on Web map layouts, is to respect the established rules of map elements in the visual hierarchy. A well-designed visual hierarchy, emphasizing data and map elements that are crucial for message communication, remains important.

Key Concepts

- People read print maps; they use Web maps. Always design Web maps with the map user in mind.
- WIMP interfaces are pointer based (e.g., they use a mouse). WIMP interactivity needs to be accommodated on most Web maps, although it is quickly becoming less used.
- Post-WIMP interfaces take many forms and continue to evolve. They are not pointer based. Multitouch, gesture control, eye-tracking, and other types of interactivity represent post-WIMP interactivity. Creating maps that facilitate post-WIMP GUIs is increasingly important.
- Map elements have largely become interactive GUIs. As such, they remain integral to the map but are far more dynamic than they once were. They can be designed to allow the map to respond to a user's interactions.

Further Reading and Resources

Online Resource

Cartography for Swiss Higher Education. http://www.e-cartouche.ch/. This virtual campus offers an online course in multimedia cartography, including extensive resources on Web map interactivity. In addition, many resources and lessons on map interactivity are provided.

Further Reading

Roth, R. E. (2011). Interacting with maps: the science and practice of cartographic interaction. Doctoral dissertation, Pennsylvania State University, University Park.

Roth, R. E. (2012). Cartographic interaction primitives: framework and synthesis. *Cartographic Journal, 49*(4), 20.

Roth, R. E., & Harrower, M. (2008). Addressing map interface usability: learning from the Lakeshore Nature Preserve interactive map. *Cartographic Perspectives, 60*, 46–66.

3

Map Elements

Introduction

Map elements are the composite parts of a map in its entirety. Consensus on definitions and categorization is somewhat rare when it comes to cartography and mapmakers; however, largely the elements found in Figure 3.1 have traditionally been considered some of the core map elements.

As becomes obvious looking at Figure 3.1, Web maps employ additional elements that are not on this list, and some of the elements in Figure 3.1 are frequently omitted on Web maps. However, the decision to include or exclude these elements on your Web maps remains the same regardless of the medium: to help communicate or promote your map's purpose.

So, what has changed between print and Web map elements? Many things. First, there are many new Web map elements that can be used (e.g., multimedia graphics or embedded videos). Second, most Web map elements are now best thought of as GUIs (graphical user interfaces). That is, almost every element can, and often should, be interactive to some extent. In the next section, we look at Web map elements one by one, reviewing how an element's design potential has changed in Web cartography.

Title/Splash Screen

One thing that has begun to disappear in Web maps, particularly mobile map apps, is the title. This makes sense in some circumstances. For example, when a Web page has the title of the map embedded in it already and the map is merely a component of the page as a whole, titles are rarely necessary. This is similar to how print maps in books tend to omit a title in favor of a caption. Another example is with a map app. Having the title of the map on the screen is often redundant since the user has to select the app from an icon and has, presumably, found and downloaded the map

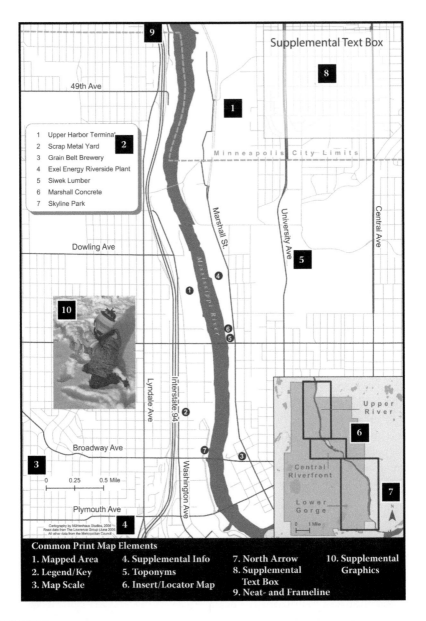

FIGURE 3.1
Print map elements.

app from a mobile store with a thorough description and screen captures of what the map app includes. Screen real estate (as is discussed in Chapter 4) is precious on mobile devices, and taking up a chunk of the screen to remind people what they are looking at via the title might be deemed a waste of space.

However, there is a tendency these days to skip a title altogether in Web maps. This is often a terrible idea. The first problem is that when you omit a title, people may not truly know what information they are viewing. A second issue is that titles are useful tools for framing your map's argument and providing useful cues regarding how people should use your map. Titles help inform a map user "what" they should be looking for. They act as cognitive cues. Without such cues, people's thoughts and interpretations are more likely to wander. Finally, titles are partially ornamental and can help make a Web map seem less disposable, less ephemeral, and more permanent. When done well, titles lend legitimacy, rhetorical power, and longevity to the map being displayed. In a society where everything on the Web is increasingly becoming disposable, well-designed titles give maps a feeling of staying power.

When it comes to mobile map apps, however, omitting titles makes more sense. There is a need to do so because people's fingers are often quite chubby (note: mine included), and showing an entire map is often done in 5 cm or less of space. However, effective map apps will employ a title splash screen. Just as almost all commercial apps begin with a splash screen showing everything from the names of developers to animated pigs grunting, map apps generally should as well. Again, title splash screens promote the map as a production, not something to be merely used once and discarded. They make a map more legitimate. Finally, as with Web map titles, brief splash screens showing a title and some poignant aesthetics can help guide the map user's interaction with the forthcoming map. In some cases, even your Web maps may benefit from short title splash screens, particularly if the API (application programming interface) you are using makes it difficult to incorporate a well-designed title into the map itself. An additional benefit of an exciting intro is that it provides great eye candy while your Web map is loading data in the background.

If you are going to use either a title or a splash screen, some simple rules should be followed. First, keep them short and to the point. Second, when possible, let the map user click on the splash screen to skip it. Regular users of your map will probably end up clicking on the splash screen hundreds of times in vain if you do not provide this option (similar to people who hit the "Close Door" button in modern elevators). Third, the splash screen should be used as a modern map title; it is nothing more, nothing less. Most people opening a map app are not going to be interested in any details about who designed the map, data sources, ellipsoids, or base map providers. A splash screen should consist of the title and perhaps a few supplemental graphics that inform users what the map is going to be about—nothing else. Supplemental information can be found elsewhere in the map (further discussed later in this chapter). Never forget that just as with well-designed traditional titles, splash screens are meant to influence the emotions one brings to the map viewing and help guide what type of information is going to be communicated.

The Argument for Splash Screens

Many information architects are certain to disagree with me on this topic. They may argue, quite rightly in many cases, that splash screens are a waste of a map user's time. They will argue that splash screens prevent people from getting to the information they want as directly and quickly as they can. This is true when it comes to Web sites. Maps are not Web sites, however. The effective use of a splash screen will depend on the goals of the map. Information architects often believe that the goal of information design is to present information, or data, clearly and truthfully to an audience. Thus, it could be argued that if you are designing a map that is going to be used simply for finding the closest pizzeria, a splash screen is completely unnecessary and annoying. As a bona fide pizza addict, I would concur with this.

However, many maps are not made simply to present raw data. Maps are often used to frame information (i.e., make an argument) and communicate knowledge (i.e., show evidence for an argument) or do both of these things. Frequently, the goal of a map is to guide, or inform, a map user's opinion about a topic. (I realize it might be unpopular to make this argument, but it is true. Government planners, for example, do not design maps of their community development plans in hopes people will see information they completely disagree with and demand the plan be entirely rewritten. They design maps to sell the public on their, it is hoped, well-informed and thought-out plan.) Adding a title splash screen often helps bolster users' confidence in the map and information they are using.

One question you might ask is, "Why has the prominent display of titles withered with Web mapping?" It may be partially due to the early limitations of APIs and HTML. Originally, when many of these APIs were developed, it was presumed these maps would be embedded in Web pages that would allow for the design of titles. Also, contemporary information architects are into minimalism. Having read Edward Tufte's books (see "Further Reading" at the end of this chapter), many may opine that flamboyant titles and splash screens are typically "chart junk." In many cases, they would be right—not always, however. Figure 3.2 highlights a good use of titles in Web maps and map apps.

Mapped Area

Explaining a mapped area used to be a straightforward endeavor. A mapmaker chose a scale at which to present data, ensuring the area of importance filled up most of the paper, and then fit the other map elements around or over it. One always wanted to emphasize the area being mapped (i.e., the specific area at the heart of the communication), making it as large as possible. Actually, this

FIGURE 3.2

This is an excellent example of the aesthetic and rhetorical power of using titles in your Web maps. It helps transform this map from what could have been another run-of-the-mill Web map into a piece of art. This map is mentioned again several times throughout the book. (Base map: 2013 MapBox, http://www.mapbox.com. Map design: CartoDB, http://www.cartodb.com).

description of mapped areas still holds true for Web maps. Except for a few minor things—like the fact that scale does not have to be fixed—the mapped area can often be manipulated by the map user, and mobile maps almost always need to have a mapped area that takes up the entire page or screen.

The fact that map scale is no longer fixed is no small issue; mapped areas can now be designed to be interactive, and many, if not most, map users expect them to be. Since the advent of Google Maps in February 2005, "slippy maps" (i.e., maps that one can interact with to pan to areas) have become a new cartographic standard (see Figure 3.3). These maps typically also include a zoom element. A single Web map might be comprised of 20 static base maps, each generalized for a specific zoom level.

Unfortunately, people often want what is not necessarily good for them. Not all maps require one to zoom in on the data for a clearer overview. In fact, the communication goal of many maps is to give someone a broad overview of general patterns and tendencies within the data. Looking at data in detail may actually mislead or misrepresent the message being visualized. An example of this would be a map showing state-based data. As Figure 3.3 demonstrates, zooming in does nothing to help understand the data being shown; it makes the map unintelligible.

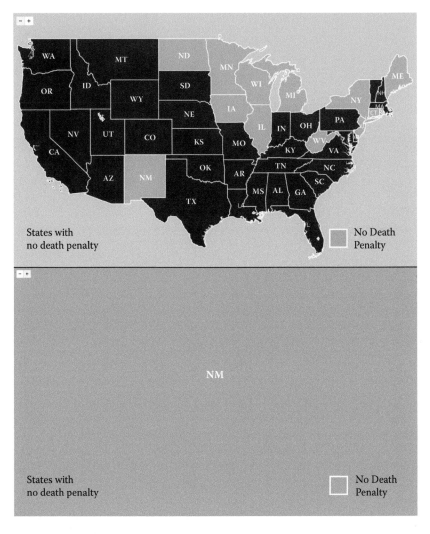

FIGURE 3.3

Top: A generic, unstyled slippy map highlighting states without the death penalty in purple. Zoom tools can be useful when the data are complex and transcalar (i.e., become more or less detailed at different scales). In this example, all zooming allows a map user to do is become confused and potentially lost within your map. Bottom: Look—New Mexico is purple.

Nonetheless, slippy maps afford numerous benefits over fixed-scale maps and, in most nonthematic circumstances, are indeed truly beneficial. The three key elements found in slippy maps, although not always together, are panning, zooming, and map rotation. These interactive techniques typically require GUIs for point-and-click devices (e.g., computers). Thus, these three things are outlined next as separate map elements based on one's need to include these elements in a map. However, never forget that if any one of these elements does not help your map users receive your message, it should probably be left out of the final product. Not all maps need to be slippy.

Pan User Interfaces

Panning is the ability to move a portion of visible content out of the mapped area to be replaced by adjacent, but currently unseen, content. If map users can zoom in on your map, or if your map is preset to a scale that does not allow the entire mapped area to be shown on a single screen at one time, your map will require the ability to be panned. There are several keys to effective panning. First, do not let map users pan beyond the scope of your mapped area (i.e., if you are mapping France's wine regions, do not let map users pan to Belarus). Panning to places unknown can result in map user confusion, and it will indubitably have an adverse impact on your communication goals. Second, do not include panning if it truly does not add anything. If you can show all of your data clearly and effectively at one scale, and everything fits on the screen, skip it. Just because you can include a map element does not mean you should.

Panning is typically achieved using several types of interaction and is almost always both WIMP (windows, icons, menus, pointer) and post-WIMP compliant regardless of the device used. The most common methods of panning are via the use of (1) panning arrows, (2) arrow keys, and (3) click and dragging. In some cases, you have two- or three-finger panning as well, but as will be discussed, in reality this is simply a more nuanced variation of click and dragging.

Panning Arrows Should Be Banned

Panning arrows are a remnant of Web maps pre-dating slippy maps. Numerous online map companies existed before 2005. However, most of these companies did not embrace the use of click-and-drag panning. Instead, they used panning arrows. When someone used panning arrows, the area next to the currently viewed area on a map would be loaded. So, if you clicked the west arrow, the area to the left of what you were currently viewing would be loaded. Panning arrows still work in much the same way today, although the data loading has changed dramatically and now typically appears seamless.

Panning arrows are clunky to say the least. They are intuitive, but they require taking your attention away from the map to find the exact arrow you want, then clicking on it to move the map. Only then can you move your attention back to the mapped area, at which time you need to reacquaint yourself with the data. These arrows are a cognitive nuisance. Moreover, they are almost universally unnecessary because today almost every interactive map has a click-and-drag feature or keyboard shortcuts that allow you to pan a map when necessary.

Yet, panning arrows are included on many Web maps. To me, they are somewhat akin to north arrows on print maps. They are included by convention, even when they do not add any value to the map. Some people will always believe that every map needs a north arrow. Some people will believe that every Web map needs panning arrows. I do not. These are a remnant of earlier technology limitations. Unless you have a really good reason to use panning arrows, I recommend excluding them. They take up a lot of screen space, most people never use them, and when people do use them, they are cognitively disruptive to the interpretation of your map.

Arrow Keys

One would think that if panning arrows are redundant and too archaic to still be used, the arrow keys found on standard QWERTY keyboards probably are as well. A multitude of mobile devices do not even have physical keyboards, much less arrow keys. However, arrow keys are extremely different from, and far superior to, panning arrows. In fact, whenever possible, I recommend incorporating arrow keys into one's map design when panning is required. However, there is a key caveat to this: make sure arrow keys are *not the only method* of panning available.

Arrow keys are available on almost every keyboard-enabled computer device. So, if your map user is interacting with a WIMP interface, the user will probably have arrow keys. If the user does not have arrow keys, alternative options will be provided (see the next section). However, the use of arrow keys to pan a map provides several benefits over the seemingly more advanced click-and-dragging method detailed in the next section. First, the majority of map users will be able to use both hands. Arrow keys allow map users to use one hand to control panning via keyboard while interacting with the map and the map data using a pointing device. Thus, two tools are allowed for map interaction, ameliorating the need to switch tasks while reading the map simply to pan it a bit. Second, arrow keys allow for interface redundancy. That is, it is always a good idea to provide map users choices about how they interact with a map and what they can do with it. Third, while arrow keys are redundant and old school, they do not take up any screen real estate or require any GUI design. They are just there if someone wants to use them. They are just an option, and map users love options.

Click and Drag and Touch Pan

If your map has pan capability, most map users will expect it to have click-and-drag panning or, for mobile touch devices, touch-and-drag panning capabilities. Click-and-drag panning is by far the most intuitive method of panning available as it simulates moving a paper map across a physical desktop in front of the map user. The best things about click-and-drag panning are that it works with both WIMP and post-WIMP interfaces and typically does not require any variations in programming to make it work on both interfaces.

Although intuitive, there are some drawbacks to click-and-drag. Fortunately, these can easily be remedied. Depending on how data rich your map is, there is a chance that it may be difficult for map users to find a spot to click before dragging that does not result in some other type of accidental map interaction, such as bringing up an information window. One work-around is to make panning a multitouch gesture. For example, in some interfaces you can move windows around the desktop or pan a map by pressing and dragging with multiple fingers (typically two or three). In sum, almost every map that has panning as an option should include the click-and-drag panning method of interactivity. However, before including panning on your map, it is advisable to make sure your map needs it to communicate effectively and that your data will not interrupt the panning navigation.

Zoom User Interfaces

I would be remiss if I did not mention that people love the ability to zoom in on maps. However, if the goal of your maps is to allow people to see changes broadly, adding zoom functionality is not advisable. One suggestion is to limit the amount of zoom possible on your map so that users feel like they have the choice to zoom but cannot actually miss the map's message when using this option. For example, if one were mapping the number of heart attack deaths by province in Germany, perhaps adding a zoom bar would make map users feel more comfortable with the map. However, one could limit the amount of zoom so that no matter what zoom level the map user chose, most of Germany was still visible in the mapped area. Better yet, avoid using a zoom unless you have detailed data that cannot be seen at a fixed scale.

In addition to zoom user interfaces, there are numerous other conventions for allowing map users to interact and change a map's scale, including zoom bars, double clicking and tapping, scroll wheels, pinch-to-zoom interactions, and keyboard shortcuts. These conventions stand out mostly because people tend to notice and react negatively when a map with a zoom feature does not use at least some of them.

Zoom Bars

The most ubiquitous interface design for zoom change is via a zoom bar (see Figure 3.4). Zoom bars take many guises, but the reason they are ubiquitous

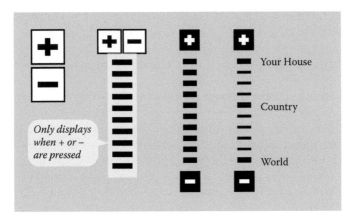

FIGURE 3.4
Examples of different zoom tools and designs.

is because they work with almost every type of user-computer interactive tool (e.g., mouse, touch pads, touch screens). Not all zoom bars are equally effective. What follows is some practical advice for zoom bars.

First, most zoom bars are vertical in nature. This design probably helps reinforce the idea of moving higher from or closer to a traditional map. In almost all cases, the top of the zoom bar is the maximum map scale (i.e., more detail, less geographic area covered, less generalization). The bottom of the zoom bar is the minimum map scale. This has become car-tographic convention and should be adhered to whenever using a vertical zoom bar. It is also convention to place it to the left or right side of the map so that it is readily available when desired but not competing with the data on the map.

Sometimes, however, the most effective zoom bars are horizontal in nature; it depends on the nature of a map's layout and its intended aesthetic appear-ance. In many nonmapping programs, horizontal zoom bars are found in a menu toward the bottom of the application (e.g., in word processors or design packages). This also may be a good place to put your zoom bar (in the lower part of your map) as people are creatures of habit. If that is where they are used to finding zoom bars in other programs, they will probably look there on your map first for a horizontal zoom bar. The left side of the horizontal zoom bar should be the smallest map scale (i.e., less detail, more geographic area covered, more generalization) and the right side the maxi-mum map scale.

Akin to scale bars on print maps, less is often more. The less gaudy zoom bars look, the less distracting they are when someone is reading the map. Keep them simple and refined. Another option that is becoming increasingly common is to have zoom bars that are not visible at all times but appear when you select plus-minus zoom buttons (discussed in the next section).

This allows for the devotion of more screen space to the map itself while allowing for easy zoom bar access when necessary.

Plus and Minus Zoom Buttons

Like north arrows, the inclusion of plus and minus zoom buttons is probably due more to cartographic norms and conventions than need. Unlike north arrows, however, pluses and minuses often take up even more screen space. I actually encourage people to avoid using plus and minus zoom interface designs whenever the amount of space is limited for the map itself (which is almost always the case on mobile devices). If you want to include plus and minus symbols on your Web map, the best method is to incorporate them as part of a zoom bar (as several examples do in Figure 3.4).

Plus and minus zoom symbols can enhance usability. For example, if someone wants to zoom in at fixed intervals (one step at a time), plus and minus buttons may be nice. However, a zoom bar allows the same incremental zooming, so having both would be redundant. Also, plus and minus zoom buttons may be useful if someone suffers from all three of the following technological shortcomings: (1) no mouse with scrolling capability, (2) no physical keyboard, and (3) no multitouch on his or her device. I am not sure how common this is, but I have my suspicions it is rare.

Double Click and Tap

Within the past handful of years, double clicking with a mouse (or double tapping via touch input) has become a new zoom standard. It is the norm not only in mapping but also in Web browsers. Thus, adding double-tap zoom functionality to one's own Web maps is a wise decision. However, you do not want map users to rely only on double clicking as it can be difficult to figure out how to zoom back to the original scale. Adding another zoom technique to your Web map is therefore warranted.

Scroll Wheel and Pinch to Zoom

Both scroll wheels and pinch to zoom are methods of using one or two fingers to zoom both in and out on a map without any clicking or tapping. Several best practices exist for scroll wheels and pinch-to-zoom interfaces. First, the zoom interaction should always emphasize the area where the user initializes the interface interaction. For example, if someone is looking at a map of Europe and initializes a scroll or pinch-to-zoom action over Aarhus, Denmark, the area around Aarhus should be enlarged at the expense of the rest of Europe (see Figure 3.5 for an example). Aarhus, and most of Denmark broadly, should then start filling the screen. Fortunately, most APIs, should you be using one, have this function built in. However, if you are designing your own custom map from scratch, this zoom behavior is important to remember.

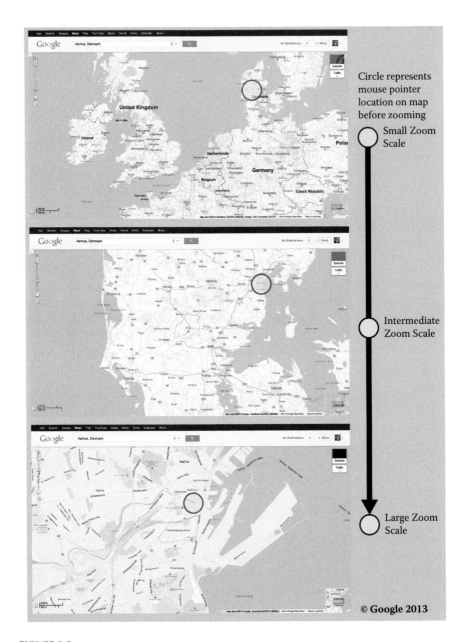

FIGURE 3.5

Zooming in on Aarhus, Denmark, with Google Maps. This is an example of exemplar zooming. If someone uses a scroll wheel as the zoom interface, the zooming should not occur on the map broadly but should center on the location of the pointer device. The same applies to pinch-to-zoom interfaces. (Map Copyright 2013 Google. Used with permission.)

Keyboard Shortcuts

Keyboard shortcuts are always easy to include (since they generally require minimal scripting), and they can enhance one's map dramatically for those using a conventional personal computer. I already discussed the use of arrow keys for panning. All of the benefits that apply to including arrow keys as an option for panning also apply to the use of the plus (+) and minus (–) key for zooming. By allowing map users the keyboard zoom option, they can use their second hand on a touch screen, touch pad, or pointing device to interact with your map in more meaningful ways. Also, their attention is never shifted away from a map's information to find a zoom interface. The one downside of keyboard shortcuts is nonexpert users will likely never realize they are there.

Rectangle Zoom

The last common interactive zoom device is the rectangle drawing tool. This tool can be useful on WIMP-based devices (i.e., computers using a traditional mouse). The tool is usually represented via an icon that the user selects. This turns the pointer device into a magnifying tool that allows the user to click and drag over part of the map area to zoom in on a particular specific region of the map. It gives the map user more precise control over where and how much zoom should be used.

This tool is useful. It allows users to skip intermediate scales and go directly to the desired one. This has the added benefit of speeding up the map zooming process on computers with slower Internet connections. Plus, users love having tools that give them the feeling of total control over a map. The problem is with post-WIMP devices (particularly multitouch ones). The demise of map interaction via pointer device not only has diminished the functionality of rectangle zoom tools but also has made them go virtually extinct on Web maps.

Several drawbacks of this tool are as follows: Clicking and dragging with your finger on a touch device is almost always interpreted first and foremost as a panning gesture. Having users select a zoom rectangle tool before clicking and dragging is possible but somewhat clumsy from a user interface standpoint. What if the user forgets to deselect it and wants to pan immediately after zooming? It will zoom in again. Moreover, with increased access to high-speed Internet, particularly on mobile devices, the need to worry about skipping scales for loading speed is greatly diminished. Finally, on multitouch devices, pinch to zoom does the same thing more intuitively than the rectangle box would. In the end, if you have space for this zoom tool and you believe most people will be using your map on a personal computer, there is no harm including it. However, omitting this tool will probably not cause too much irritation on the part of map users as long as you include scroll, pinch to zoom, keyboard, or other alternative zooming methods.

Zoom User Interfaces Conclusion

There is no dearth of GUI and interactive options for including zoom on your Web maps. As always, keep in mind several things before you decide which methods to include. First, what is the communicative purpose of your map? Is zoom necessary? If not, will allowing map users to zoom in on your map hurt the potential efficacy of your message? Can you allow zooming to facilitate people's love of slippy maps but limit the map's slipperiness so that the message is not disregarded or missed entirely?

Second, on which type of device is your map most likely to be accessed? WIMP-only devices (unlikely), post-WIMP-only devices (unlikely), or both (very likely) are the choices. You should choose methods that support both types of devices, but you should rarely include all of the methods. In fact, many Web maps these days seem to include far too many interface options for zooming at the expense of the map itself. Keep it simple.

One way to keep it simple is to compare and contrast the pros and cons of the different methods (see Figure 3.6). Keeping in mind your communicative goals and your audience, you can choose several of the most ideal methods for incorporating zoom into your map. You can also combine or tweak existing methods. For example, the plus and minus buttons have largely been incorporated into zoom bars these days. Some maps now merely have a button that you click that then opens a variety of zoom features; this way, zoom tools do not take up precious screen space. Play around with the zoom features, but remember, some of the norms are now convention, and people do not like to learn new interface designs just to suit your needs.

Pros and Cons of Different Zoom Tools		
Tool	**Benefits**	**Drawbacks**
Standard Zoom Bar	Easy to use Can see which zoom level you are on	Takes up a lot of space Rarely used with mouse scroll and pinch-to-zoom Low accuracy zooming (zooms center of mapped area)
+/− Buttons	Easy to use Use minimal screen space	Cannot see which zoom level you are on Rarely used with mouse scroll and pinch-to-zoom Low accuracy zooming (zooms center of mapped area)
Mouse Scroll Wheel	Easy to use No screen space taken High accuracy zooming over features	Cannot see which zoom level you are on Can easily zoom in or out too far and get confused
Pinch-to-Zoom	Easy to use No screen space taken Some accuracy zooming over features	Can easily zoom in or out too far and get confused Can vary in accuracy depending on screen and finger sizes Cannot see which zoom level you are on
+/− Keys	Easy to use No screen space taken	Few people realize these work Can easily zoom in or out too far and get confused Cannot see which zoom level you are on

FIGURE 3.6
Pros and cons of different map zoom tools.

Map Rotation Interfaces

Map rotation tools are now found in many Web maps. In some cases, interfaces allowing a map user to rotate a map are crucial. In most cases, though, such interfaces are simply not needed. Again, if a map element is not needed to help communicate your map's information, do not include it.

One thing to keep in mind when designing for map rotation is that different devices will have different rotation interaction capabilities. On WIMP interface devices, you will need to design a GUI that can be dragged with a mouse. However, for multitouch and most post-WIMP devices, you will find that a GUI is not needed at all—although a graphic that highlights the orientation of the map is always needed when north is not at the top. It is best practice to design for both types of devices.

Two Types of Rotation

There are two types of rotation that are possible: automated and user controlled. Automated rotation is when a map user has no control over the rotation; it occurs no matter what. This is typically found in narrative maps that guide map users through a timeline. It can also occur in animated portions of maps, flybys for example. Alternatively, user-controlled animation is that which is powered by a GUI. This interface tool will allow map users to flip the map horizontally and often vertically.

Regardless of which method of rotation you choose, *always* include a graphic element that shows which direction the map is currently facing. If the map is changing directions frequently, then your graphic element will need to be animated and change direction with the map itself. If your map user has control over the rotation via a map element tool, then it is good practice to incorporate this map orientation element into the tool itself (see Figure 3.7).

Several best practices should be adhered to when including rotation in your Web map:

1. *Always start with north at the top.* This is a convention that has not changed with Web mapping as the majority of people assume north is at the top of the map.
2. *Make rotations smooth.* Rotations should be smoothly animated. They need to be intuitive and graceful. Jagged rotations can discombobulate a map user and make it difficult to know which direction one is going.
3. *Never present text upside down.* If the map rotates, the text also should rotate. Make sure that your map reader does not have to read upside-down text.

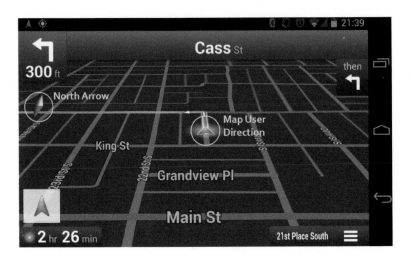

FIGURE 3.7

An example of a north arrow facing the direction of a map user as the device is rotated and used for navigation. The map faces the direction the user is facing. (Copyright 2013 Google Maps. With permission.)

4. *Maps should rotate to mimic the direction a person is facing.* On a mobile device map that is designed for navigation, it is good practice to have the map rotate so that the direction one is facing is at the top. North should always be at the top when a map is opened, but the map can then rotate to face the direction the top of the device is pointing. Again, *never forget* to include a compass-type graphic so that people know which direction they are actually looking, walking, or driving (see Figure 3.7 for an example).

Information Window Design

Information windows (info windows for short) are best thought of as callout labels on steroids. Map users expect these on Web maps; they have become a new map element in and of themselves. They can be large or small in size. They can be information dense, embedding everything from text, hyperlinks, images, videos, music—the options are endless. As usual, though, endless options are not always a good thing.

Many info windows are poorly designed because they include either too much or too little information (see Figure 3.8). Finding the right balance for communication can be a challenge. It is recommended that information not directly pertinent to the goals of the map should generally be excluded from the information window. Instead, within the info window

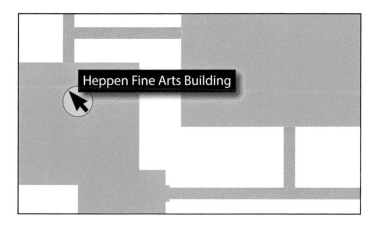

FIGURE 3.8

Example of tool tips. Tool tips are information that pops up in a very small window as a pointer device moves over an object. Generally, this is used to provide a name or label to an item that is not labeled on the base map and does not have enough information for an info window.

create hyperlinks to supplemental content so that someone can access it if needed. Conversely, if you do not have any useful information about a spatial dataset, simple map labels are always more effective than animated info windows presenting someone with nothing more than a place name. It is nearly always a waste of time and energy for someone to have to click to get a name. Maps have had place names on them for millennia; info windows are not required.

If you do not have the ability to add labels to the base map, and you do not have enough information for info windows, I recommend using tool tips (Figure 3.8). Tool tips appear when someone rolls over an object in your dataset with a pointer device. A small, simple text box appears to describe what the person just pointed at. However, with post-WIMP devices, the rollover effect typically does not exist. Include it on the base map when possible.

Info windows are incredible tools. They allow us to show or connect to far more information than was ever possible with a static map. They eliminate the need for supplemental text boxes and callout labels. One serious drawback to info windows, however, is that they have the propensity to cover up an inordinate amount of mapped area. This means that once the information is retrieved, no real inferences can be made about the highlighted object and places in the vicinity around it. Therefore, it is good practice to make info windows as compact as possible and allow the map user to decide whether to expand the info window or open links from within it. Finally, always keep the info window near the selected feature.

Including media, multimedia, and even alternative GUIs within info windows is now common and can prove both informative and aesthetically stunning. One thing to keep in mind when embedding an image or video is that these are best placed at the top or in the middle of an info window. When placed at the bottom of the info window it often makes the window itself look unbalanced. Figure 3.9 shows several examples of well-designed info windows that are balanced and include interesting GUIs within them.

One thing to consider is whether you want, or need, connector lines (or connecting arrows) between the info windows and user-selected places at all. Most APIs include arrows by default. Arrows can be very useful for map users so that they know they selected what they intended to select. However, they also add one more graphic element that covers the mapped area. These arrows tend to work well with point symbols, but when someone is selecting a large polygon, the object the arrow is pointing at may not be as clear. For the most part, arrows are expected these days, but it may be worth considering using an alternative method of highlighting the object to which an info window refers. Finally, arrows are simply not an option for those who are using side panels instead of info windows to display data and information. Several alternative methods of highlighting a place that was selected by a map user include changing the color and size of the selected object or animating it briefly. This allows a map user to look back and forth between the information panel and the mapped area and quickly find the object of interest.

Another important consideration is to make your info windows touch friendly. An info window is inherently touch *unfriendly* if the only way to

FIGURE 3.9

Several examples of well-designed info windows. Info windows should be aesthetically pleasing and blend in with the style of the rest of the map. Left: Notice how the embedded photo does not dominate the window or map, and if a map user wants more info, the user can simply click a button. This allows the info window to remain small. (Map by University Communications and Marketing, University of Wisconsin–Madison. Used with permission.) On the right, notice how the info window uses subdued text values, avoids using a window outline, and minimizes the length of the arrow connecting it to the features. (Map by CartoDB using MapBox base map.) On maps that make heavy use of info windows, as much thought should be put into their design as is put into the design of the mapped area itself.

close your info window is to click on a little *x* in the upper right-hand corner. These are difficult to hit with a finger on a first (or even the fifth) try and often result in map users accidentally selecting other parts of the map.

One workaround is to include the *x* for those with computer mice but also make the info window disappear when a map user simply touches the info window. Obviously, this method will not work well if you have interactive material in your info window. Therefore, a second workaround is to have the info window disappear if someone clicks anywhere outside it. Perhaps most advisable, for the foreseeable future, is to make the close button larger (i.e., perhaps an entire side or top panel within the info window).

Finally, info windows are only as effective as they are attractive. When possible, style your info windows to match the aesthetics of your map (and, when applicable, Web site). This can be as simple as using the same type-faces found on your map in the info window; incorporating colors from your map into the graphics or type used in your info window; and using CSS (Cascading Style Sheets) to design them. Whenever possible, do not settle for a default info window. Design your windows to better highlight and reinforce your information.

Locator Maps

Locator maps are inset maps that show where the main mapped area is located within a broader geographical context. They are useful for large-scale maps (i.e., maps showing small areas) because people often have a difficult time referencing the location of what they are viewing based on their mental maps alone. Early on, most Web maps included locator maps, and they are also a good idea to include with your large-scale maps.

Locator maps have an added bonus. Not only do they help map users find where they are in the world, so to speak, they also can function as an alternative zoom tool. Many locator maps let the user drag a rectangle over them or pan the area currently being viewed by directly interacting with the locator map. This offers map users an additional method of zooming and getting to where they want to be as quickly and directly as possible.

In recent years, there has been a trend away from locator maps among online map services. Certainly, locator maps can use up valuable screen real estate; however, they can easily be made interactive so that they can be shown or hidden with the press of an interface button. Thus, when possible, it is recommended that you include a locator map if your mapped area is large in scale.

Most often, locator maps are found in the bottom right of a Web map. This is not required, but it makes sense from a map-reading standpoint, as it falls in line with the map user's natural gaze across the map (from upper left to lower right). Sometimes, locator maps are incorporated into interactive legends—to be discussed later in this chapter—or in conjunction with zoom and rotation tools. There is no true standard yet. The main goal is to keep the locator map low in the visual hierarchy (which is discussed in Chapter 4). Do not let it dominate. It is best to have it hidden or let the map user hide it if he or she does not want to use it. (This power can be given to the user by simply creating a minimize/maximize button for the locator map.) Figure 3.10 shows an example of a locator map.

Menu Design

Something that one did not have to worry about with static maps was the design of menus: drop-down, horizontal, vertical, icon-based, text-based, interactive menus. Makers of paper maps simply placed the important information in supplemental text boxes or in the legend. Today, map users expect to have options. How many options should be available to a person is up to the mapmaker to determine, but menus are a great way of organizing and hiding many of them so they do not interfere with map reading. Menu styles and layouts are as myriad as map types. However, there are certain rules to menu design that people should be aware of before designing one of their own.

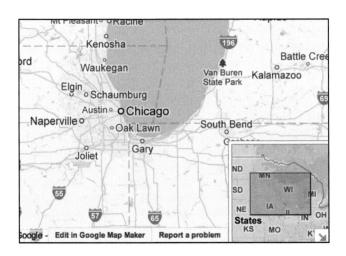

FIGURE 3.10
An example of a locator map. (Copyright 2013 Google.) Allowing a map user to open and close a locator map as desired is ideal, as the map element tends to take up valuable screen real estate and is not always needed. (See Chapter 4 for more on screen real estate and visual hierarchy.)

The Three-Click Rule Is Wrong But It Remains a Useful Axiom

The three-click rule has mythical status in the world of Web design. The rule states that a Web user, and in this case a Web map user, should never have to click more than three times with a pointing device (e.g., mouse) to obtain the information desired. This rule rests on the belief that people like to find things on the Web quickly, or they will become frustrated and leave a Web site. It is true that people become frustrated and leave Web sites when they cannot intuitively find what they are looking for. However, it is absolutely untrue that the number of clicks one has to make to find information has anything to do with this frustration (Weinschenk, 2011).

Organize, Test, and Confirm

User studies continually show that the organization of your menus is far more important than how deeply information is hidden within a menu system. The key is that you want map users to know instinctively where to look for information they are seeking. If they are on the right track, they do not care how many clicks it takes. On the other hand, if they have trouble finding a tool and they start clicking on different menu items in vain, they will become frustrated and start to feel upset with the interface and the map. So, how do you figure out how to organize your menus for effective user interaction? The answer is to use user experience testing.

Step One: Organize Your Menus

The first step is to take all of the functionality of your map—anything that has an interactive component that is not directly addressed by a map element—and organize these things into groups. One great way to do this is to take a stack of sticky notes or index cards, and on each one jot down an interactive or menu item feature you plan to include on your map. Lay them out and start grouping these menu items into categories based on which ones you would expect to find near one another in a toolbar. For example, perhaps you want to include a print feature, a share via social network option, and a save a PNG (Portable Network Graphics) option as several menu items. These three features will probably fit nicely into one category as they are all related to map distribution.

Every map will have slightly different menu options based on the purpose of the map. So, this organization game would be beneficial with each map project. For other categorical organizing tips, I highly recommend the book *Gamestorming* (Gray, Brown, & Macanufo, 2010). I also recommend having others not involved with the map design provide input on how they would categorize the menu items. It is always a good idea to bring in an outside perspective.

One other thing to take into consideration when organizing your menu elements is established convention. There are already many conventions about where people expect to find things in menus. For example, you expect to find the options to turn on/off certain map layers, such as traffic, points of interest, and the like, in one drop-down menu or side panel. In general, it is advisable to go with conventional norms when using standard menus. However, whenever designing your own menus, you will have to organize them in meaningful ways, and categorization exercises may help.

Step Two: Design Your Menus

Once the interactive and functional components of your Web map are categorized and organized into groups, the next step is to design your menus. There really are no guidelines regarding how to do this. The main point is to make your menus blend in with the map so that they are readily accessible and intuitive and do not take too much attention away from the map. They should be styled to fit aesthetically with your map—similar colors and typefaces help.

For example, if you are designing a Web map that will most likely be used on mobile phones, design the menus using button interfaces that reflect the nature of a given mobile device's operating system. Developers and designers have created numerous skins—interface styles—that mimic the look and style of a variety of operating systems (e.g., Apple iOS and Google Android). Many of these are freely available for download and incorporation into your Web maps. Use these to your advantage to help make your map feel like an integral part of a device or computer. Default menu buttons are often bulky and generic looking; taking the time to tweak your interface to fit the communication objectives of your map and the device your map is being used on will likely make your map more intuitive and fun to use.

There is some debate about what works best for menu headings: icons or text. In general, it is often believed that people can make cognitive connections with icons more quickly as they do not need to read them; also, icons take up less space than words on the screen. However, if icons are completely unique and generally unknown, then they may just slow down user interaction with your map. Designing icons is a topic well beyond the scope of this book, but keep this in mind when designing menu icons: Icon standards are rarely universal, and they are constantly changing. If space allows, evidence suggests it is advisable to use both an icon and text (Roth & Harrower, 2008). When in doubt, simply use text. However, make sure the text is succinct and descriptive. A menu heading called "Communications Hub," for example, is not only excessively long but a little too vague. Using a title like "Share" may be better.

Finally, it is always a good idea to keep the menu chrome—the area your menus take up on the Web map—to a minimum. Keep your items as compact and unobtrusive as possible. Screen space is limited, and people want to view the map, not your menus. One way to minimize menu chrome is simply to have menus appear and disappear as needed. Perhaps you can shrink all of your menus down into a simple button or along a thin pane adjacent to the edge of the map (see Figure 3.11). When a map user clicks on the button, the menus can appear. This technique of hiding menu chrome is becoming quite common in mobile device interfaces, as well as in personal computer applications.

Step Three: Conduct User Tests

The next step is to conduct user tests. This can be as simple as inviting your neighbors over for dinner and inviting them to play with a new Web map you are designing (awkward but effective). Essentially, you are going to watch people—preferably people who have never used or seen your map before— use your map. Look for areas of concern or confusion. Watch them to see if they become frustrated or if your map does not live up to their expectations at some point. Give them a task to do with the map (i.e., find a certain location or answer questions about the information they are viewing). Ask questions about why they are selecting certain menu items or, even more telling, why they are not. In essence, all you have to do is observe and collect information

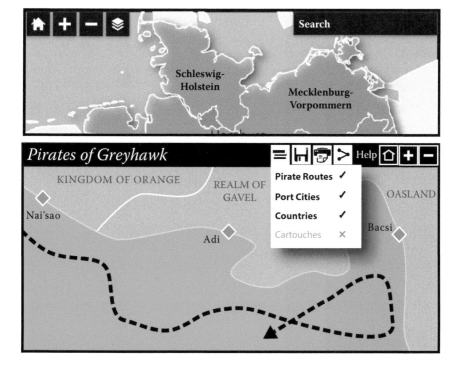

FIGURE 3.11

Examples of stellar map menu design. Compact and minimalist design is always preferred for mobile devices. However, never make your menus so compact and abstract that they become obtuse.

on how people actually use your map. For a great book on Web design and user testing, see Steve Krug's *Don't Make Me Think!* (2005). The book's citation is listed in the "Further Reading" section at the end of this chapter.

Step Four: Reorganize Your Menus

Once you get over the fact that your test subjects do not think about menus like you do, you can incorporate your findings into a revised menu system. It is hoped this merely means rearranging several of the menus into new categories that are more intuitive to most users. In some cases, though, this may mean an entire revamp of your menu system. If major changes are required, it is advisable to retest your map. It is the only way to ensure that your map will work for map users.

Supplemental Information

The inclusion of supplemental information—including cartographer names, data sources, projection used, and other useful, authoritative information about the map and its data—was once one of the last things people worried about placing on static maps. It was generally placed on the map in as inconspicuous a place as possible and in a miniscule font size. Most agreed that such information should be included, but where it was included was of minimal consequence as long as it did not detract from the map.

By and large, the consensus regarding print map supplemental information remains unchanged for Web maps. What has changed is how such information is included. No longer does this need to be directly included over any part of the map unless you are using an API or spatial data that require you to do so. In fact, supplemental information is often best written up on a separate Web page or splash screen and loaded only when desired by the map user. You can still make such information easily and readily accessible for the more technically inclined map users by including a "Help," "About," or "?" button in one of the menus itself. If the information has little relevance to the purpose of the map, never include this information on the map itself unless you have to by law.

Help Menu

A rare necessity with print maps is the inclusion of a help menu. Not everyone will understand how to interact with your Web map, though, so always, always, include a help menu item or a help button that opens up additional information for those who need assistance using your interface. You can create a Web page or pop-up help window that succinctly acts as a how-to guide. This can be done using text, but often a series of images or even a short video will be more useful.

Neat Lines and Frame Lines

Although it may seem superfluous, many Web maps, particularly those designed for embedding within Web pages, still need neat lines or frame lines. These map elements are just as important now as they were for print maps. They help separate the map from other Web page content and help people know where the map starts and ends within a page. Unfortunately, many well-designed maps on the Web no longer use neat lines and frame lines, leading to a less-intuitive interpretation of a map.

With mobile map apps, however, neat lines are largely analogous to the size of the screen itself. Since mobile devices by their very nature have smaller areas of display, the screen border itself should almost always function as the neat line. This is covered in more detail in the map layout chapter.

Designing Smart Legends

One key map element has not been mentioned thus far. It is often the first map element people think of, and it is often the most important: the map legend (also known as map key). Legends are almost always a crucial map element. Legends are central to the effective communication of any complex visual information. Yet, how legends are constructed has changed so dramatically, and Web mapping legends can take so many different forms, that the topic has earned its own master section in this chapter.

The Purpose of a Legend Remains the Same

The goal of any map legend, regardless of medium, is the same: to act as the Rosetta stone for your map. Not all maps need legends because some maps are designed so simplistically and intuitively that including a legend is redundant. However, almost every map showing detailed data of any sort needs one, as otherwise people have to guess what it is you are representing and attempting to communicate. If people are making guesses, informed or otherwise, the effectiveness of your communication is impaired. And as we know, nothing should be designed for or omitted from a map if it might impair its communication.

Web map legends often fall short of their sole reason for existence in one of two ways. They either fail to include enough information for a map user to make sense of what the user is looking at or they include far too many irrelevant interactive options that distract the map user from paying attention to the information found on the map.

How to Avoid Creating Ineffective Web Map Legends

First things first: design with a purpose. All you need to do to design effective Web map legends is follow three simple steps. First, decide what your legend needs to show—not what it could show (e.g., water is blue), not what it might show (e.g., every map symbol found in a national map database), but what it needs to show (e.g., the values of your thematic color scheme). Second, determine how to display these data clearly. Make sure that your legend is not an enigma to users. Space is not so much of an issue as you can easily make legends minimize to a panel or icon in most cases. Third, decide whether the effectiveness of your communication will be enhanced or diminished by allowing the map user to interact with the legend. Legend interactivity is the focus of the next section, but I would like to offer a word of caution here. Interactive legends can be useful and powerful and really make playing with your map fun for an end user. Most often, Web map legends benefit from some interactivity (e.g., the ability to turn some data layers on or off), but too much of a good thing can be a bad thing. Careful consideration should be given to what potential impact this user interactivity will have on your communication goals. Too much interactivity, particularly activity that potentially minimizes the clear communication of your information, can be quite detrimental.

How Interactivity Makes Legends More Powerful

Legend interactivity can help map users make a map more intuitive and clearer. Making your legend interactive offers users the capability to transform aspects of the map based on individual needs. Reviewing all of the types of interactivity and GUIs that could be created for legends would be a book in and of itself. Instead, I review several rationales regarding why you would want to allow users to interact with your map legend.

Having Options = Feeling in Control

The ability to click buttons and turn things on and off makes people feel good. Moreover, interacting with legends can be really fun, particularly if the legends are well designed. If your map is fun, people are more likely to remember it and, it is hoped, get some meaning and use from it.

Even having your legend simply open and close with the click of a button or tap of the screen may be enough interaction to excite users. In fact, generally it is a good idea to always let users minimize and maximize your legend. It is rare that someone constantly needs to refer to a legend while looking at a map. However, having it conveniently accessible empowers map users. It lets them decide if and when they need to refer to it.

Layer Interactivity

In many cases, legends are a great place to provide the opportunity to highlight or add more data to a map in the form of layers. There may be times that your map user may need to turn different types of data on or off, reorder data in the hierarchy, add aerial photography, add a topographic base map, or import data of their own to place on the map. Organizing Web map legends by data layers like this is becoming increasingly common. Just make sure the additional data help your map achieve its purpose. Several examples of this are included in Figure 3.12.

Map and Data Generalization

Contrarily, many times maps confuse users by providing far too much information. My mother, bless her heart, can never find her way down to my house and back (a 2-hour drive each direction) using her dashboard GPS

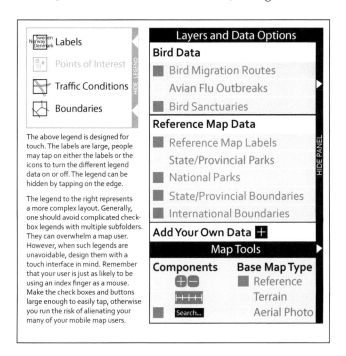

FIGURE 3.12
Examples of legend layout and style.

or any of the map apps on her smartphone. She becomes overwhelmed and confused by the maze of streets on these maps. Drawing a simple sketch map is far more effective for her and is an example of when less is more for map users. Providing tools in your legends to minimize base map data or emphasize selected routes and datasets is a great idea. Of course, it depends on the purpose of one's map. I wish for my mom's sake that map service providers more frequently had an option to include far fewer data on their maps instead of more.

Symbol and Thematic Modification

Perhaps no place lends itself better to allowing a map user to select, change, or manipulate how your map data are represented than the legend. While you have chosen and designed your symbology and data representations based on an overarching map purpose, sometimes including options for the manipulation of your map elements helps map users style the map as they prefer. If you believe that providing map users these options will not hurt your map's purpose, it is recommended to add a few interactive tools that modify thematic styling within your legend.

There are a variety of map components that can be potentially manipulated through a map legend. Symbols can be changed or resized; polygons or objects can be recolored; toponyms can be resized or turned on and off. Again, be careful to make sure your map does not devolve into a cornucopia of miscellaneous interactivity that has nothing to do with the purpose of your map. Simple menus and choices are often best (see Figure 3.13).

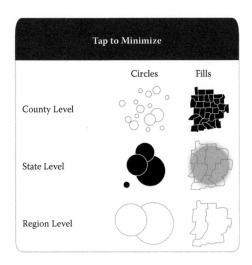

FIGURE 3.13

An example of a simple user interface allowing someone with chubby fingers to quickly change the thematic visualization of the data. The purple circle represents someone's fingertip to provide scale.

Temporal Legends for Animations

Temporal legends are essential whenever your map includes animation. Interactive timeline design is covered in detail in Chapter 10. One thing that is worth noting here given this chapter's emphasis on interactivity, however, is that these legends are often most effective when they are interactive. People want to have control over animations: the ability to play and pause, at minimum, and ideally slow the animation down or speed it up.

Also, if the animation is central to your map's purpose, keeping this legend on screen at all times likely makes sense. Animation timelines should be prominent in the visual hierarchy so that a map user can readily refer to them while watching an animation and thereby know how to accurately interpret the animation. For more information on timelines, be sure to read Chapter 10 where a complete overview is provided.

Other Web Map Elements

There are many elements of dynamic and Web maps that do not neatly fit into the aforementioned categories. These elements can be tools used to interpret and interact with map data on the map itself or supplemental interfaces found within the mapped area that enhance the map's communication. Most of these things can be created as independent graphics or are embedded within other map elements (e.g., menus, map area, legends).

Multimedia Graphics

Online mapping *is* multimedia. One's maps are no longer limited to merely presenting static abstractions of reality; they now present dynamic abstractions of varying detail. For example, a simple map feature that we take for granted these days—the ability to switch the base map from a standard cartographic representation to aerial imagery—is an example of multimedia. You are switching from an artistic visualization to a data snapshot. Add to this the ability to add myriad images (e.g., Google Street View, Yahoo! Flickr, Panoramio, and so on) and easily embed videos (e.g., YouTube, Vimeo), and suddenly maps are a cornucopia of media.

Two types of multimedia are most frequently used in Web maps: images and videos. In this section, I review the effective use of these two forms of media in your maps and discuss some potential pitfalls to avoid when incorporating them into your maps.

Images

Images, including photographs, vector drawings, or artwork, are often ideal supplements to Web maps. They can either add information to particular variables on the map or can be used to enhance the rhetorical and emotive power of your message. There is just one rule of thumb when using images on your Web maps: Do not let the images dominate the mapped area and spatial data.

The most common way to avoid this problem is to embed any images dealing with different data points or polygons into info windows. Info windows, as already discussed in this chapter, can be spruced up dramatically with a couple of photographs or pieces of vector art. In most circumstances, though, you want to make sure the info window itself does not just become a glorified picture frame. When possible, show a thumbnail or small size of an image within an info window and then allow the map user to click on the thumbnail to open up a larger version of that image (Figure 3.14).

When your images are directly over your mapped area or appear somewhere other than in an info window, it is often a good idea to provide users the option to hide the images. Obviously, this should not be done if the images are integral to your communication. However, providing users of your map the option to disable rollover images (e.g., country flags

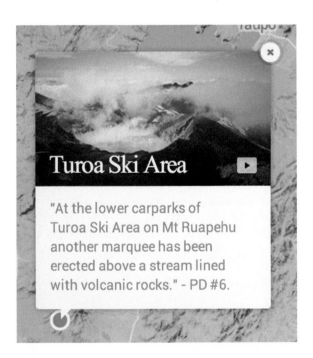

FIGURE 3.14
An example of a thumbnail that opens up a YouTube video when clicked. Notice that a little play button icon is put over the picture so that the map user knows it will open a video of some sort. (Copyright CartoDB. Used with permission. Base map: MapBox. Map available at: http://cartodb.github.io/cartodb.js/examples/TheHobbitLocations/.)

that pop up on the screen when people roll over a country on your map with a mouse) is generally a good idea and gives map users the feeling of empowerment.

Videos

Many of the same caveats for images also apply to videos. These are best embedded in info windows. In the info windows, videos are ideally styled to integrate with the information being shared. On your map itself, videos are best displayed at small sizes so they can be played within an info window. Even better, users might have the option of clicking on a thumbnail of a video to load the video in a separate Web browser window. (This is easily done using an online video streaming service like Vimeo or YouTube.) As with images on your map, be careful to avoid letting videos consume the entire mapped area. You can either create a pop-up window for the video to play in or size the video accordingly so that it does not cover the entire map.

Videos should rarely be set to automatic play (i.e., as soon as a user clicks on an object, a video appears and begins playing). Typically, you want to leave it up to the map user to decide whether to view a video at a particular time. There may be circumstances when this does not apply (e.g., maybe you are showing traffic conditions when people click on a webcam), but generally, let the map user decide when to view your videos.

Graphs and Highlighting

One of the greatest things about Web mapping, particularly given the ability to import real-time data from a plethora of databases, is that we can now link interactive graphs and charts directly to our maps. Many Web maps will never have a need for using interactive graphs and charts (or what I call "chart graphics"). However, whenever you are mapping quantitative thematic data (as discussed in Chapter 9), adding charts to your map can reinforce your information as well as make it rhetorically effective (i.e., numbers with charts appear authoritative).

Information graphics are often useful but can also clutter a map's layout quickly. Depending on the purpose and style of your chart graphics, the following guidelines for implementing their design are recommended: Place the chart graphics in a side or bottom panel. Allow users to turn these chart graphics on or off or retract the panel they are in to free screen space when necessary. This is particularly important when designing for small-screened mobile devices. For mobile devices, instead of using a panel for your chart graphics, you might consider having a separate menu screen. On this screen, users might interact with your chart, highlight attributes that they would like to see mapped spatially, and then go back to the map window to view the map.

Avoid putting interactive chart graphics in info windows, at least ones that link directly back to the map itself. Doing so can result in some serious problems. If someone wants to highlight some data in the chart graphic to see where they appear on the map, it may be difficult for someone to see what is highlighted with an info window sitting on top of the mapped area.

Tips for Effective Map–Chart Graphic Highlighting

Highlighting and brushing are effective interactive processes for showing where data in a chart are found on a map. Highlighting is when you select an area of a map with a pointer device to highlight that area's corresponding data in chart graphics. Vice versa, one can select information on a chart to highlight an area on the map. Brushing refers to rolling over data with a pointer device to bring up information, which will disappear when you move the pointer away from that particular location. Although highlighting and brushing are extremely useful tools for data exploration, many times they can be quite distracting for maps that are simply meant to present information. This is another feature, like chart graphics, that works best when you allow map users to opt-in and do not set it up as a default.

Key Concepts

- Do not use default map elements without some map-specific styling or placement. Whenever the opportunity arises, stylize them yourself. Default map elements make your map less memorable, more difficult to discern from millions of other Web maps, and potentially less aesthetically pleasing.

- Map elements should be included or excluded based on two things only: the purpose of your map and your intended map audience.

- If a map element does not facilitate the communication of your message, omit it.

- If your map audience is expecting particular map elements due to conventions or norms with which they are familiar (e.g., those pesky panning arrows) and leaving these elements off your map may have an impact on the success of your communication, consider including said elements. Caveat: Do not use this rule as an excuse to include all of the map elements available to you.

- Map titles are still important and should always be included in some capacity. They can be extremely effective for guiding map user interpretation of your map.

Further Reading and Resources

Web Sites

Developer.com. The theory behind user interface design, part one. http://www.developer.com/design/article.php/1545991/The-Theory-Behind-User-Interface-Design-Part-One.htm (short URL: http://goo.gl/1SKJk). This is a brief yet quite thorough introduction to user interface design.

Fifty Photoshop interface design tutorials. http://www.antsmagazine.com/photo-shop-2/tutorials/50-photoshop-interface-design-tutorials/ (short URL: http://goo.gl/7MZW3). This is a list of Photoshop tutorials, all dealing with interface design. Most of these Photoshop designs can be refashioned into graphical user interface devices for your Web maps (e.g., info windows or zoom bars). If nothing else, perusing this list of tutorials will give you inspiration. If you do not have Photoshop, many of these tutorials can be completed using GIMP (http://www.gimp.org/), a free, open-source photo-editing program.

Further Reading

Dent, B. D. (1999). *Cartography: thematic map design* (4th ed., p. 434). Dubuque, IA: WCB.

Gray, D., Brown, S., & Macanufo, J. (2010). *Gamestorming: a playbook for innovators, rule-breakers, and changemakers* (Google e-book) (p. 286). O'Reilly Media. Retrieved from http://books.google.com/books?hl=en&lr=&id=_-xnEDNPxwYC&pgis=1

Krug, S. (2005). *Don't make me think: a common sense approach to Web usability* (2nd ed.). Berkeley, CA: New Riders.

Norman, D. (2002). *The design of everyday things*. New York: Basic Books.

Roth, R. E., & Harrower, M. (2008). Addressing map interface usability: learning from the Lakeshore Nature Preserve interactive map. *Cartographic Perspectives, 2*(60), 46–66.

Shneiderman, B., & Ben, S. (1998). *Designing the user interface*. Delhi: Pearson Education India.

Tufte, E. R. (1983). *The visual display of quantatative information*. Cheshire, CT: Graphics Press.

Tufte, E. R. (1991). *Envisioning Information*. Cheshire, CT: Graphics Press.

Weinschenk, S. M. (2011). *100 things every designer needs to know about people*. Berkeley, CA: New Riders.

4

Map Composition and Layout

The change from paper to an interactive medium has implications for map design far beyond map elements. It has altered how maps can be designed and organized. The purpose of this chapter is to get you thinking about online Web map composition and layout. Map composition deals primarily with a map's visual hierarchy; it is the process of deciding which map elements, and components within the map area, to emphasize and promote over others to best communicate your message to map users. Web map layout is intrinsically tied to decisions you make about your map's composition. However, layout specifically deals with the arrangement and balance of map elements on the screen. A good way to keep the two straight is that map composition deals with "how" prominently to display different map elements; map layout deals with "where" to display them.

The rest of this chapter covers a variety of core concepts resulting in Web map design that is more intelligent and aesthetically pleasing. First, the concept of visual hierarchy is reviewed and modified to best reflect the nature of Web mapping. Map layout is discussed next. Here, we review three elements that have a direct impact on the layout of every single Web map: screen real estate, screen resolution, and pixels per inch (ppi). Then, a discussion of the benefits and drawbacks of fluid versus compartmentalized map layouts for Web maps is conducted. Finally, key questions to ask yourself as your Web map layout is conceived, designed, and produced are reviewed.

Map Composition and Visual Hierarchy

I have always been a fan of Borden Dent's (Dent et al., 2008) visual hierarchy for map design. He defined visual hierarchy as the placement of all map objects and elements into a logical order by their relative importance (Dent et al., 2008). It is my goal in this section to reinvigorate the idea that, depending on your map's purpose, certain map elements should be emphasized over others, and some elements should be excluded altogether.

Dent noted that the importance of a map element depends on your map's communicative purpose. Although each map is a unique communiqué requiring consideration over which elements should be emphasized for a particular audience, Dent proposed a general visual hierarchy to guide cartographers of reference and thematic maps (Figure 4.1). His hierarchy was based on what he called intellectual levels (how important elements were to a map's successful communication) and visual levels (how much something had to be emphasized).

Dent's visual hierarchy is effective for designing meaningful maps in the print medium. I still use it to teach my students in introductory cartography. However, in many ways, it also represents a bygone era when maps were exquisitely designed layer by layer, with effort being put into the design of almost every base map, including detailed, manual labeling, and the placement of specific map elements like graticules. Today, many Web and print cartographers do not even design their own base maps. They often borrow other people's, or more often corporations', base map tiles and map their own information over them. Thus, they have minimal control over the visual hierarchy of their base maps.

Dent's Visual Levels for Thematic Maps	
Level 1a	Thematic Symbols
Level 1b	Title Legend(s) Map Symbols Labels
Level 2	Base Map: Land Areas Political Boundaries Significant Physical Features
Level 2–3	Explanatory Materials: Map Sources Credits
Level 3	Base Map: Water Features
Level 4	Other Base Map Elements: Labels Grids Scales

Adapted from Cartography, Borden Dent, 1999: p. 252

FIGURE 4.1

Borden Dent's visual levels for thematic maps. This hierarchy is meant to guide map designers in determining what map elements to emphasize when designing thematic maps for print. (Modified from Dent, B. D., Torguson, J., & Hodler, T. [2008]. *Cartography: thematic map design.* New York: McGraw-Hill, p. 252.)

Another concern regarding Dent's visual hierarchy is that the map elements in use have changed dramatically. When Dent proposed his model, every map element had to be placed on a static page with a mapped area that was set at a fixed scale. Today, as already discussed extensively, map elements can be minimized or hidden or pop out from a side panel at the desire of the map user. Thus, not all of the elements need to compete concurrently with one another for a Web map user's attention as they did in print. Still, a Web mapper needs to know how to make certain map elements stand out compared to less-important ones and how to design elements hidden in menus so that they are easy to find.

Techniques of Emphasizing Map Elements in the Visual Hierarchy

The best way to make a map element stand out in the visual hierarchy is to design it so that it contrasts with other map elements. Elements that contrast with background or neighboring map elements will be brought to the fore of the visual hierarchy. There are innumerable methods to establish contrast. Some of the simplest are to manipulate color hue, value, and intensity (color is discussed in more detail in Chapter 5). Styling your objects so that they look different from surrounding objects (e.g., giving your legend a sharp-edged rectangle appearance when other objects use rounded rectangles) can also work, although you should be weary of losing an overall aesthetic by varying the styles among map elements too much.

Conversely, to make a map element better blend into the lower levels of the visual hierarchy, one can design minimal contrast between it and its surroundings. Grouping a map element with other items and matching their appearance is one way to do this. Objects tend to be grouped in one's field of vision when they are near one another, of the same shape, and of the same size (Ware, 2008). Thus, menu items are often grouped together perceptually by a map user into a single "menu" object. The individual buttons of a menu are not discerned unless a user visually focuses on the menu to explore for a particular button. Also, layers represented in an interactive legend will not necessarily be individually discerned, but rather they will be grouped together perceptually until the map user decides he or she would like to select or manipulate a particular layer. Designing map elements so that they are similar in size and appearance and are near one another is key to simplifying your map's visual hierarchy.

New Visual Hierarchies for Web Map Elements

This section outlines three new visual hierarchies for Web map design (Figure 4.2). These are based on Dent's original hierarchy and my own observations. Just as with Dent's hierarchy, these new Web-specific visual hierarchies can *and must* be manipulated depending on the communicative purpose of your map and the intended audience. These hierarchies are

VISUAL HIERARCHY LEVELS FOR WEB MAP DESIGN					
General Interest Web Maps		**Thematic Web Maps**		**Animated Web Maps**	
Level 1	Title/Splash Screen Map Symbology Key Reference Data Info Windows (opened)	**Level 1**	Title/Splash Screen Thematic Visualization Legend	**Level 1**	Title/Splash Screen Animation Symbology Map Symbology Temporal Legend/Interface
Level 2	Base Map Base Map Labels Navigation/Directions Tools	**Level 2**	Base Map (generalized) Info Windows (opened) Chart Graphics	**Level 2**	Base Map Legend Info Windows (opened) Locator Map
Level 3	Map Interactivity Pan/Zoom/Rotation Tools Print/Share Map Features	**Level 3**	Base Map Labels Map Interactivity Pan/Zoom/Rotation Tools Menus with Additional Tools	**Level 3**	Base Map Labels Map Interactivity Pan/Zoom/Rotation Tools Menus with Additional Tools
Level 4	Locator Maps Chart Graphics Multimedia Supplements	**Level 4**	Locator Maps Multimedia Supplements	**Level 4**	Multimedia Supplements Chart Graphics
Level 5	Supplemental Information Attribution and Copyright Neatlines/Grids/Graticules Tool Tips	**Level 5**	Supplemental Information Attribution and Copyright Neatlines/Grids/Graticules Tool Tips	**Level 5**	Supplemental Information Attribution and Copyright Neatlines/Grids/Graticules Tool Tips

FIGURE 4.2

Three visual hierarchies for Web maps. The first is for reference and general-purpose maps, such as a map highlighting a local community's amenities for potential visitors. The second is for thematic maps on the Web (covered in Chapter 9). The third is for animated maps on the Web (covered in Chapter 10).

merely meant as a guide in deciding which map elements are crucial for your Web map and how prominently these should be displayed on your map. One thing to keep in mind is that many of the elements discussed previously in this book can be made almost invisible by burying them behind buttons and menu items. This is a design advantage print cartographers did not have.

Reference/General-Interest Web Map Visual Hierarchy

The main goal of most reference maps is to facilitate map user knowledge about locations, distances, and directions between places. General-interest maps are defined here as maps that show general information designed specifically to communicate an argument or story to an intended map audience. For example, this might be a map highlighting the testimonials of people who will be positively impacted by a piece of national legislation. The biggest difference between the two is that map users choose which data to look for on a reference map—it is a top-down searching process. General-interest maps are designed with specific information highlighted and emphasize that information to the map user.

Thematic Web Maps

The main goal of thematic maps and data visualizations is to communicate information in a convincing manner about a topic or several topics. One of the most effective ways of communicating is via the title. Thus, a title on your map or an introductory splash screen should definitely be designed to catch the map user's attention. Obviously, another core aspect of your map is the thematic visualization, or visualizations, themselves. These should be bold and effectively stand out from the base map. Legends are often crucial for interpreting and understanding data; thus, they should generally be emphasized more in thematic Web maps than in reference ones. Menus and navigation bars are typically less important in these types of maps. However, their position in the hierarchy may need to vary depending on how much user control you allow over the thematic data. When included, interactive chart graphics should be emphasized, but rarely at the expense of the map representation itself. The base map is typically only meant to communicate the geographic context of the thematic data; thus, base maps that are more minimalist and simplified are most appropriate.

Temporal Animated Web Maps

The main goal of a temporal animation is to effectively highlight the distribution and movement of a process over time. Effectively designed temporal animation maps generally have a similar visual hierarchy to thematic maps except that they also must emphasize the movement and diffusion of an element. They should also prominently display an animation legend.

Map Layout and Organization

The purpose of map layout is quite simply to organize the map elements in a coherent and intuitive package to convey your information effectively to map users. The entire goal of map layout is organization. (Thus, throughout the rest of this chapter I refer to map layout and map organization interchangeably.) Yet, there is remarkably little discussion in layperson English on best practices for online map layout. Most people simply follow the conventions (i.e., defaults) of the application programming interface (API) they are using or mimic the appearance of run-of-the-mill map services in their own custom-made Web maps.

The obstacles confronting print map layout are very different from those that Web maps face. Print cartographers can typically plan and design their map elements based on how large the final print map will be. A print cartographer must also determine at which resolution a map will be printed, although given the nature of the medium, this rarely has an impact on a map's layout.

Web map designers are not afforded these luxuries. One cannot design a Web map for viewing at a fixed screen size. In fact, it is impossible to know at which resolution a Web map will be used or how large your map will appear on different devices. Moreover, as discussed in previous chapters, most map elements have now evolved into interactive graphical user interfaces (GUIs), which means they do not necessarily take the same form, amount of space, or style as their print predecessors—or even between different Web maps.

Screen Real Estate, Resolution, and the Pixel Problem

Three things will inevitably affect every Web map's layout: (1) screen real estate, (2) screen resolution, and (3) pixels per inch.

Screen Real Estate

Screen real estate refers to the amount of space you have to work with on any screened device. You have a fixed amount of space to show everything you want to show. This is not so different from what print cartographers had to deal with when it came to paper maps. One advantage Web mappers have over those designing print maps, however, is that they can often include a zoom interface (as mentioned in Chapter 3) so that data can be shown on a single map at multiple scales instead of at one fixed scale (as on paper).

However, there are also several major hurdles to deal with as a Web cartographer with regard to screen real estate. First, unlike with a print map, the Web cartographer never knows how large a screen the map user will have. The screen may vary in size from an 80-cm HDTV (high-definition television) to a 3-cm MP3 player. Second, often Web mappers have to design their maps to be embedded within other applications (e.g., Web browser, mobile operating system), many of which will be sized differently, have different size chromes, and vary substantially in style depending on the device used to view the map. (*Chrome* is the amount of space an application uses for its own elements, e.g., on a Web browser the space reserved for the window tabs; forward, back, home, and refresh buttons; address bar; search bar; bookmark bar; and footer area.) A map viewed in one browser using a certain operating system may have far less space than a map using a different browser in a different system. Likewise, a Web map application viewed on a smartphone from 3 years ago may have far less screen real estate at its disposal than the same map on a more recent smartphone (also known as a "phablet" for approaching the size of a tablet). Figure 4.3 highlights the impact of screen size on screen real estate.

Before you begin any project, you want to consider all of the different screen sizes your Web map is likely to be viewed on and design accordingly. You may have to design the same map several times to facilitate functionality in different browsers or for different size mobile devices. Never settle for a one-size-fits-all Web map layout. Whenever possible, include code in your HTML that will allow your map to determine the resolution and screen size of the user's device. Then, have an appropriately laid out map display. (Visit

FIGURE 4.3

Screen size and real estate have an impact on how much space you have to show your map. Not taking this into consideration can result in maps that are difficult for people to use and aesthetically subpar. Here, two different size smartphones are shown with the same map. Notice that much less space is left for the mapped area on the smaller phone when using the same map elements. Also, notice the difference in finger distance should one want to pinch and zoom over this area because the map has been resized to fit both screens.

this book's accompanying Web site at http://www.ian.muehlenhaus.com/ webcartography/ for links to tutorials on how to include code like this in your Web site.)

One rule of thumb is always to design your map for the device with the smallest screen you want people to use your map on. Please note the qualifier: You should not plan for the device with the smallest screen *possible*, but the smallest screen that *you are willing* to let your map users use. Part of communicating your Web map effectively is limiting how you will let users view the map. Although you have no control over the wide array of devices on which users may attempt to view your map, you certainly can design for a minimum screen size.

The benefits of designing for the smallest suitable screen size are that you ensure usability on the smallest devices you want people to use and that you use your limited screen real estate judiciously. Moreover, even though some devices may have to upscale your map to fit their gargantuan screens, indubitably resulting in pixelated map elements, the elements and interfaces will still function and be usable. That is the most important thing.

Screen Resolution

Screen resolution is a pretty straightforward concept. It is most simply defined as how many vertical columns and horizontal rows of pixels a screen can show. Resolution is normally represented using the number of horizontal rows, but different resolution standards also have a variety of acronyms. For example, high definition (HD) is also referred to as HD 720 because its screen resolution is 1280 × 720 pixels. Full high definition (FHD) is referred to as HD 1080 because these screens are comprised of 1920 × 1080 pixels. Figure 4.4 highlights the difference between the two HD formats.

Fifteen to twenty years ago, the concept of a screen's resolution, or what I also interchangeably refer to as display resolution, was a relatively simple thing to deal with. Most monitors were cathode-ray tubes (CRTs) and had by today's standard's a very limited resolution. Standard monitors were 640 × 480 or 800 × 600 pixels, nearly all with a 4:3 aspect ratio. (Aspect ratio is simply the fraction created by dividing the number of vertical columns by

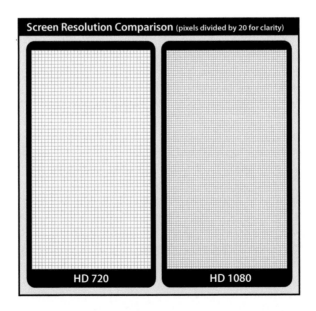

FIGURE 4.4
The difference in screen resolution between two high-definition formats. To make this figure legible, each pixel here represents 20 pixels on both devices.

horizontal rows.) You were supposed to design your maps to have resolutions less than these standards to ensure most people could actually view your maps on their screens without having to use scroll bars.

Major advancements in technology, as well as increasingly varied aspect ratios, have resulted in different types of screen resolutions being distributed on all sorts of devices. The laptop this is being written on has a screen resolution of 2048 × 1536; however, I am currently looking at a monitor that has a display of 1920 × 1080. Meanwhile, when I hook my laptop up to a projector in a classroom, the information is typically displayed at an 800 × 600 resolution.

The takeaway is that you cannot plan your design for the screen resolution of a generic user's device as easily as you once could. Even on the same device, depending on how a person decides to view your map (e.g., on the native screen, via external monitor, etc.), it can be viewed at multiple resolutions. It is therefore good practice to design your map with a resolution that will fit nicely on a majority of screens without the need for scrolling. The ideal resolution to design for is continually moving upward, so I dare not make a suggestion in print. Instead, at the end of this chapter I recommend several Web sites that provide usage statistics on which type of screen resolutions are being used most frequently to browse the Internet. I also offer several recommendations for free resources that will allow you to test how your maps look at different resolutions right inside your Web browser.

Pixels Per Inch

Pixels per inch represent how many pixels are shown in a line 1 inch in length on a given screen. (This concept is sometimes erroneously, though harmlessly, interchanged with dots per inch. Dots per inch refer to printed materials. Pixels per inch refer to any measurements taken on screens.) The more pixels there are in an inch, the more detail can be shown.

Thus, a higher pixels-per-inch value means a higher resolution, right? No—one thing that can be quite confusing is that screen resolution and pixels per inch are not the same thing. The phone in your pocket has more pixels per inch than an expensive flat-screen television, even though the television has a much higher resolution. For example, if your smartphone has an HD 720 resolution (720 × 1280 pixels) and your phone has a screen that is 4.5 inches in size, then it has 326 ppi. In other words, your phone is capable of extremely high image quality. (Once you get above 300 ppi or so on a small screen, it becomes impossible for the human eye to differentiate between pixels.)

Now, perhaps you are fortunate enough to also own a 50-inch, FHD resolution LCD screen. (Hey, if you design maps for a living, you might as well design them in style from a sofa, right?) FHD resolution is 1920 × 1080 pixels. However, this resolution is spaced out over 50 inches. Your humongous monitor has a better screen resolution than your phone, but it also has a measly 44 ppi. Your LCD monitor still offers an incredible image. In fact, when you sit about 2.5 m from the screen, you might swear it looks just as good as the image on your phone. It probably does. The further away your eyes are from a screen, the less the pixels per inch matter as distance diminishes the eye's ability to discern pixel density.

PPI Advancement = Big Mess

At first, it may seem like concerning yourself with pixels per inch should not matter as much as resolution. That was true a decade ago, when almost all screens were 72 or 96 ppi. Unfortunately, today the number of pixels per inch a screen displays is far less predictable, and almost every mobile device has a slightly different pixel density. The reason this matters so much is that raster images are sized by pixel (e.g., a photo might be 392 × 512 pixels). Therefore, the pixel density displayed on a screen will have a direct impact on the size of the image. High-density pixel displays are great when it comes to visual clarity; they are a nightmare when it comes to designing images, or maps, that you would like to display similarly regardless of the device.

This is particularly important when it comes to tile mapping (i.e., maps comprised of previously created PNG or JPEG tiles, such as those used by OpenStreetMap and other Web mapping services). The higher a screen's pixels per inch, the smaller the tiles will look. If you have labels on these tiles, the labels will shrink with the tiles and often become illegible. Zooming in on a tile map using a zoom bar will not help. Each tile at each scale is 256 × 256 pixels in dimension. On a television screen with 44 ppi, that would be almost 15 cm per tile. On a high-resolution, small-screen mobile device, a map tile is probably 0.8 cm if you are lucky (see Figure 4.5). Try finding map details and reading small labels on that.

Making a long story short, Web cartographers now must deal with more than just planning to design a map for a particular resolution. We also have to take into account the varying display capabilities of different devices. Whenever possible, you will want to make sure that your Web map layout can be reformatted on the fly depending on the type of screen, resolution, or device your map user is viewing. Determining screen size and resolution can be done using a variety of means, including via scripting in your Web map or on the site in which your map is embedded. Often, map APIs include options for determining screen sizes, resolutions, and devices that will automatically reorganize your map's layout. Selecting an API with this option is something to consider, as designing a one-size-fits-all Web map is rarely a good idea. This is particularly true for multitouch devices, as when a crucial map element (e.g., an interactive legend) becomes too small to touch on a given screen, your map immediately becomes unusable.

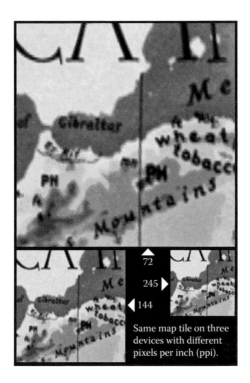

Same map tile on three devices with different pixels per inch (ppi).

FIGURE 4.5

Pixels per inch (ppi) affect how large-raster images will appear on a device. This map tile is sized proportionately for three different devices. Devices with lower pixels per inch make raster images (such as tiled maps) appear larger. The 72-ppi image represents how large this tile would appear on a standard CRT monitor; the 144-ppi image is how large the same tile would appear on a MacBook Pro Retina Display (circa 2013); and the 245-ppi image is the size it would be on many high-end smartphones.

Two Types of Web Map Layouts

Before we can begin designing Web map layouts, it is imperative that we look back at how effective print maps were traditionally designed. From there, we can see what aspects of static map layout are still relevant and which ones might be better discarded when it comes to Web mapping. Once we have determined what does not work for Web maps, we can start analyzing how to re-create or enhance certain print layout effects that do work using Web technologies. Print map layouts have taken numerous forms over the centuries, but modern layouts can be predominantly broken down into two categories: fluid and fragmented. Few maps fall entirely into one extreme or another, yet even fewer maps are a perfect balance between the two; most fall more toward one style of layout over the other.

Fluid Map Layout

Fluid layouts are those in which the neat and frame lines are one and the same (see Figure 4.6). The map takes up the entire page. All of the other map elements are placed on top of the mapped area and, when designed well, do not cover any important spatial data relevant to clear communication. Typically, the elements are scattered around the mapped area so that visual balance is achieved.

The fluid map layout has several advantages. It allows the mapped area to be produced at the largest scale possible, providing more visual detail and requiring less generalization. Moreover, it allows for a homogeneous design or aesthetic as the map elements are not compartmentalized and can be designed to match both one another and the visualization on the map. One potential disadvantage, however, is that this type of map layout can start to look very busy. If not designed well, it can result in an unbalanced map that detracts from the information being communicated.

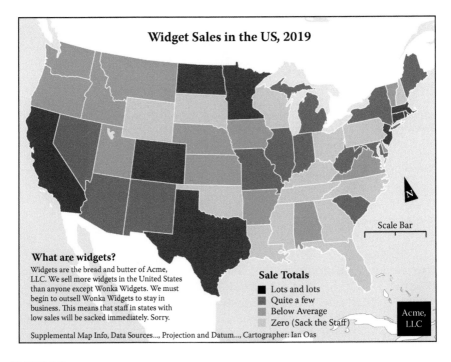

FIGURE 4.6

The fluid print map layout uses the entire map as a canvas. The mapped area covers everything. Map elements are sprinkled around the important areas of the map in a manner that facilitates map balance. The north arrow is only included as an example map element; they are generally not needed when north is at the top and should not be used on conic projections.

Compartmentalized Map Layout

Compartmentalized map layouts are those that separate many of the core map elements into separate boxes or divisions (see Figure 4.7). Each map element has its own particular place; mingling among elements is discouraged. For example, the title of a map may be in a separately framed box at the top of the mapped area itself, with its legend in a separate, boxed area underneath the mapped area. Scale bars may not actually be over the mapped area but instead adjacent to it. Supplemental text boxes might be put to the side of the mapped area.

This fragmented layout style is commonly used among geographic information system (GIS) professionals, and several variations are often included as map design templates in GIS packages. Although this method of map layout often results in polished and authoritative-looking maps, it also

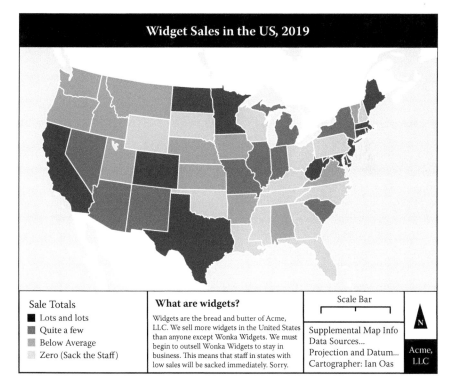

FIGURE 4.7
The compartmentalized, or fragmented, layout is characterized by a separation of map elements from mapped area. It is generally designed in a manner that clearly delineates between different elements, particularly between the mapped area and everything else. The mapped area ends up taking less space because of this. The United States here is smaller than that in the fluid layout because it is confined to a smaller rectangle. The north arrow is only included as an example map element.

suffers from several disadvantages. First, the mapped area is often not as large as it could be, limiting the amount of data detail that can be revealed. Second, fragmenting map elements means that map users may have to move their eyes much further away from the mapped area to read the supplemental information. Of course, as an advantage, once a map user becomes adequately acquainted with the layout, the user no longer needs to search for map elements; he or she will know right where to look for information. In sum, compartmentalized map layouts do organize map elements fairly well and intuitively, but often at the expense of the mapped area and, potentially, aesthetics.

Web Map Layouts

Both fluid and compartmentalized maps have continued their layout hegemony in Web and interactive mapping. Both have their advantages and disadvantages in different situations. Next, I outline how each is used in Web mapping, then I review some generic layout templates that may help you begin thinking about layout ideas for your maps.

Compartmentalized Map Layouts

Compartmentalized maps have become ubiquitous in maps designed for Web browsers. These maps frequently make use of HTML divisions (<div> tags) that are styled with Cascading Style Sheets (CSS). A frame with 100% height (going from the top to the bottom of the browser window) is often appended to a Web map on the left- or right-hand side (see Figure 4.8). This mimics a trend that started with early map services (e.g., Google Maps), whereby the side of a map is a control center of sorts—a place where search inquiries can be made, legends shown, info windows sequestered, and other map elements presented. This norm even pre-dates Web mapping, reflecting a trend that began in the 1990s, when Web sites were commonly designed using frames and tables.

Compartmentalized Web maps function well when there is ample screen real estate. However, when screen space becomes more limited (e.g., on phones), using a fragmented layout is often not a great choice. Most of the time, you want people looking at and focusing on the map, not supplemental information and legends. This is important to remember if you are designing a Web map to be used on multiple devices. You may want to design both a fragmented and a fluid version for different size screens.

Another potential issue with compartmentalized layouts is that they often lack rhetorical and aesthetic flair. Certainly, this is not always the case, and it depends on what you are attempting to communicate to an audience, but if you are not presenting formal data, using a standard, left- or right-paneled compartmentalized layout can look quite staid. Part of effective communication is exciting people about the maps they are reading.

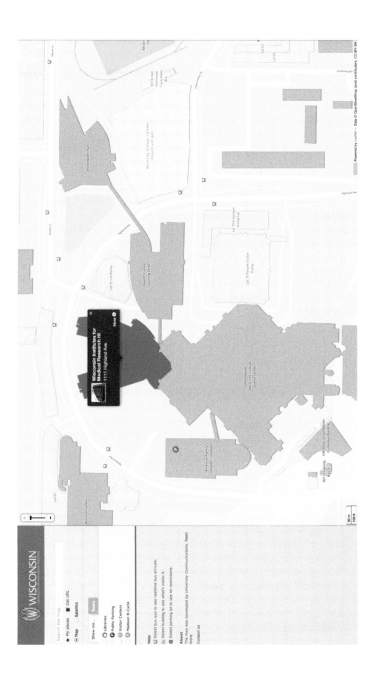

FIGURE 4.8

An example of a compartmentalized Web map using a standard left-frame design. (Used with permission from the University of Wisconsin–Madison.) The left-hand side of the map is actually a separate HTML division. To see a similar layout but with a less-fragmented look, check out Figure 3.2. In Figure 3.2, the background is the same color as the base map area. When you pan the map area, there is a CSS transparency effect so the mapped area disappears gradually. (There is a reference to a tutorial at the end of this chapter that shows you how to do this with CSS.)

Nonetheless, there are many positive reasons that the compartmentalized Web map style is ubiquitous. People are familiar with the layout. Due to their interactions with other Web maps employing this layout, most people will instantly know where to look if they want to find tools to interact with, search the map, or find a legend. Creating a fragmented map layout also lends a feeling of professional, business-like design—much akin to the fragmented design that is common on government maps. Finally, such maps are often a bit easier to design using HTML and CSS. You position your different division tags, you set their sizes, and you are done.

Fluid Map Layouts

Unsurprisingly, fluid layouts have come to proliferate on mobile devices on which screen real estate is at a premium. Fluid Web maps tend to hide features under icon-based menus and submenus (see Figure 4.9). Clicking on one of these icons will open a larger menu, legend, or tool palette within which you can manipulate or interact with different map elements. Well-designed fluid layouts have such menus collapse as soon as someone clicks outside them. (Again, the use of small *x* symbols as the only method of closing menus in the post-WIMP [windows, icons, menus, pointer], multitouch era is highly discouraged.)

The arrangement of menu icons can take many forms. However, one generally wants to place these icons where people focus their visual gaze the least (see Figure 4.9). In Western societies, this is typically in the lower left-hand or upper right-hand corner of the map. The interactive map elements remain readily accessible in these places but are not as distracting. This allows the areas people look at first and foremost when using maps to be consumed by the mapped data. Figure 4.9 provides an example of a fluid map layout.

Certain map elements simply do not lend themselves to small-screen devices and should be kept off a fluid map layout in these circumstances. Titles are often best presented as splash screens—appearing at the start and disappearing. Supplemental information is typically best hidden under a default menu option on mobile devices rather than placed over the mapped area. Panning arrows—ugh—and zoom bars are best omitted, or at least hidden and minimized, if you know a particular layout is going to be used on multitouch devices.

How to Design an Effective Web Map Layout

Thus far in the book, I have provided information about a variety of theoretical and technical issues that need to be addressed in the design of map composition and layout. What I have not yet done is give you any idea of how to incorporate

FIGURE 4.9

An example of a fluid Web map. The title and tools are placed over the mapped area. This is generally the most effective layout for mobile devices, where screen real estate is at a premium.

all of these ideas into the layout of your own map. That is the goal of the final part of this chapter. There are two principal questions that should guide you through the Web map layout process: Does the layout help you achieve your communication goals? Is the map layout intuitive? Next, I outline a five-step process to design effective layouts for all of your Web mapping projects.

1. **Identify your Web map audience and its expectations.**

   What kind of Web map does your intended audience expect?

   Knowing what your audience expects when they first view your Web map should play a big part in how you design your map's layout. If your map is going to be used in a manner that is similar to many other maps already out there, you may want to mimic the layout and design of those maps.

2. **Determine which kind of Web map you will design.**

Are you going to make a map for a Web browser or a stand-alone app?

Browser-based maps tend to be well integrated within the Web browser, allowing the browser's menus and features (e.g., printing, bookmarking, and searching) to be used on the map itself. Map apps are often stand-alone map packages, like video game apps, that have their own unique interface designs and incorporate well into the ecosystem of a given mobile operating system (e.g., Android, iOS, Windows).

Are you going to make a slippy map (e.g., using map tiles or an API), or are you going to create an interactive Web map with a static base map?

Not all Web maps need to be slippy maps. Although often advantageous when it comes to user activity and data processing, map service-based slippy maps (e.g., Google Maps API) generally put more constraints on how you can design your map layout. Designing your own static base map provides you with more design and layout options. Creating a unique layout can help your map look distinctive and exceptional.

3. **Determine which elements of the visual hierarchy need to stand out.**

Which map elements must be promoted for the map to make the most sense?

I recommend making a list of all of the map elements you *think* your map needs. Then, go through and prioritize them. Choose the top three or four. Those are the map elements that are needed. Everything else is typically supplementary.

How can these map elements be organized in a meaningful manner?

All map elements aside from the most important ones, which will typically have their own GUIs and be visible on screen to some extent, should be organized in a meaningful manner into menus or icon buttons. Group these into categories based on function and convention (e.g., the ability to print a map is usually found near the ability to e-mail the map to a friend).

4. **Design mock layouts and user test.**

Have you exhausted all of your layout options?

The best Web map layouts are not whipped together at the end of a long programming session; they are well thought out and have been compared to a litany of other potential layouts. Turn off your computer, pull out a notebook, draw a grid on the page, and start sketching every type of layout imaginable. Move your map elements and legends around, paying particular attention to the visual hierarchy you have decided on in the previous step: promoting elements that need to be emphasized, demoting other elements. It is advisable to create two sets of layout sketches: one for Web browser-based maps (WIMP compatible) and one for mobile, multitouch devices (post-WIMP compatible).

What do people think of your layout(s)?

Before you spend days and weeks finalizing a map design, it is a good idea to run the layout by several people to get their thoughts. Ideally, you will find people who fit the demographic of your intended audience. What people usually do when they look at mock layouts is ask questions. Some of these questions can be quite revealing: Why is this over here and not there?

It is always easier to fix map element and layout shortcomings before they are fully implemented.

5. Finalize the map layout and user test it again.

Are there any unforeseen issues?

For example, are there certain aspects of your ideal map layout that are impossible to implement due to the Web service, API, or technology you are using? Once you have finalized your map, it is wise to user test it to make sure it is fully functional. User testing is the best method. For more information on user testing, I highly recommend Steven Krug's (2005) very informative and entertaining book *Don't Make Me Think*.

Conclusion

Visual hierarchy and map layout remain as important as ever for map communication. If you fail to establish an effective visual hierarchy, the message or information you are trying to communicate is prone to become lost or overlooked. Visual hierarchy and map layout are central to a map's aesthetic. The aesthetic is central to retention of a message. Do not put off map layout and visual hierarchy until the last minute; plan and test layouts throughout the design process. Your message will resonate much better if you do.

Key Concepts

- Maps need to emphasize different map elements depending on their purpose. Referring to the visual hierarchies for Web maps can help you make decisions about which map elements to emphasize.
- In the era of the mobile device, screen real estate is to be coveted. Do not waste space on unimportant map elements.
- Pixels per inch and screen resolution are two related but different things. Pixels per inch will have a drastic impact on the size of raster maps (e.g., map tiles) on different screens. Take this into consideration when designing your maps.
- Both compartmentalized and fluid Web map layouts have their advantages and disadvantages. Make sure never to use a compartmentalized layout on mobile maps designed for small screens.
- When designing a map layout, remember to address the following things: audience expectation, type of map to be designed, layout design sketches, and user testing.

Further Reading and Resources

Resolution Testers

Google Chrome App Store. Resolution Test Widget. https://chrome.google.com/webstore/detail/resolution-test/idhfcdbheobinplaamokffboaccidbal?hl=en

Resolution Tester Web site. http://www.infobyip.com/testwebsiteresolution.php

Tutorial Resources

Lynda.com CSS Gradients Tutorial. This 1-hour tutorial shows you how to design all sorts of gradients, including semitransparent ones, that can help you create maps with HTML div tags that make your map look fluid instead of fragmented. I highly recommend a monthly subscription to this site in general. You can learn almost everything you will need to know about Web mapping within several weeks on this site. For a great example of using CSS gradients to minimize the fragmented look, visit http://cartodb.github.com/cartodb.js/examples/TheHobbitLocations/ (short URL: http://goo.gl/G966h).

Cartography for Swiss Higher Education: Graphical user interface—layout and design lesson. http://www.e-cartouche.ch/content_reg/cartouche/ui_access/en/text/ui_access.pdf (short URL: http://goo.gl/eI3o9). This excellent white paper includes lessons, study guides, great graphics, and many examples and is a worthy read for anyone interested in graphical user interfaces as they directly relate to maps.

Further Reading

Dent, B. D., Torguson, J., & Hodler, T. (2008). *Cartography: thematic map design* (p. 368). New York: McGraw-Hill.

Krug, S. (2005). *Don't make me think: a common sense approach to Web usability* (2nd ed.). Berkeley, CA: New Riders.

Tufte, E. R. (1983). *The visual display of quantitative information* (p. 197). Cheshire, CT: Graphics Press.

Tufte, E. R. (1991). *Envisioning information* (p. 126). Cheshire, CT: Graphics Press.

Ware, C. (2008). *Visual thinking for design* (S. Card, J. Grudin, & J. Nielsen, Eds.), *The Morgan Kaufmann Series in Interactive Technologies* (p. 197). New York: Morgan Kaufmann.

5

Color

Introduction

Selecting an effective color scheme is one of the most crucial steps in designing a powerful and useful map, regardless of whether it is for print or the Web. This chapter first reviews the three properties of color, then moves on to explore the different types of color combinations that are possible for you to use on your Web maps. I then review a few best practice rules from print cartography that stand firm. Once that is done, we dive into a refresher on several color models, including an explanation of hexadecimal versus decimal RGB (red-green-blue) color classifications. The chapter concludes with an in-depth look at things to consider when designing color schemes for base maps, reference maps, and thematic maps.

Defining and Understanding Color

Before I can go any further, we first need to define what we mean by color. I will not go into too much detail on color theory here as there are entire textbooks devoted to the topic—several of which are referenced under "Further Reading and Resources" at the end of this chapter. However, we need to have a basic understanding of how we define color before we can begin talking about which colors work well together on Web maps.

Three Properties of Color

There are three key properties to color: hue, value, and saturation.

Hue

Hue is what we colloquially refer to as "color" on a daily basis. There are pure hues that are easy to identify, including violet, blue, green, red, and yellow (see Figure 5.1). Beyond that, not many official names for hues exist. Corporations invent names that help market hues to the masses (e.g., "midnight violet," "morose gray"), but beyond about a dozen hues found in the color wheel (see discussion and figure in "Making Sense of the Color Wheel" on page 85) consensus over hue names is rare.

Value

Value is most easily defined as how light or dark a color hue is. When white is added to a color hue, it is called a tint. When black is added, it is called a shade. Hues that are lighter in appearance (e.g., yellow and green) generally result in better-appearing tint colors. Color hues that are darker in appearance to begin (blue, violet, and red) generally result in more nuanced and spectacular-looking shade colors. Playing with value expands the amount of color options we have at our disposal much more than playing with hue does. Value is an exceptional tool for intuitively showing different spatial data values, as Figure 5.2 highlights.

Saturation

Saturation refers to the brightness of a hue. Colors that are pure and unmixed (e.g., primary colors) have the highest saturation (Figure 5.3). The mixing of colors always diminishes a hue's saturation (also known as chroma or intensity). Mixing black, white, or gray with a hue has the greatest impact on its saturation.

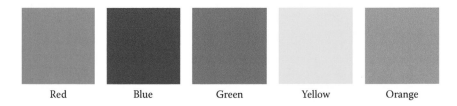

| Red | Blue | Green | Yellow | Orange |

FIGURE 5.1
Five hue examples with their commonly accepted names.

FIGURE 5.2
Example of a purple hue using five different color values.

FIGURE 5.3

Example of the color purple at different saturation values. The leftmost purple has zero saturation; the rightmost is at full saturation.

Making Sense of the Color Wheel

Everything you learned about how colors relate to one another on the color wheel in grade school remains true on the Web, except how the colors are constructed (more on that in the discussion of color models). You cannot design truly great maps without first understanding how colors work with one another. Once you have the basics down, it becomes easier to start exploring more complex and nuanced color variations that end up enhancing your communications.

Understanding color combinations begins with the color wheel as shown in Figure 5.4. From this seemingly simple wheel, a vast array of useful and

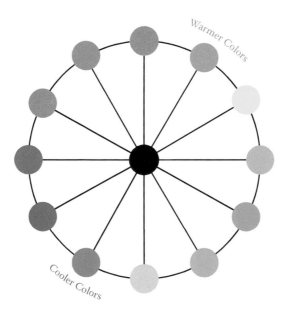

FIGURE 5.4

The RGB color wheel. Warmer colors are found in the upper-right portion of the wheel; cooler colors are in the lower left. The primary colors from the RGB color wheel are red, green, and blue. Beginning with these colors, we can create all other colors and begin to analyze harmonious color relationships.

effective color combinations can be created. Some combinations will prove most effective at eliciting emotional responses, whereas others will be better for clearly communicating quantitative differences. How you combine colors from this wheel will subconsciously determine the emotional impact and clarity of your map.

Warm and Cold Colors

Warm and cold colors refer to those colors that subconsciously infer a feeling of warmth or coolness, respectively, in those viewing the color. Due to evolutionary and environmental development, we cannot help but infer temperatures from the colors we see. In general, the colors yellow, red, and orange feel warm. When we see them, we associate them with the desert, sun, lava, fire, greasy fast-food joints, and all sorts of other warm things. The colors blue-violet, purple, green, and turquoise are cool colors. These are associated with things like water, ice, grass (a little bit warmer due to its infusion of yellows), and shade. This is important to remember, as using the wrong types of colors to map data will make your map less intuitive for audiences to interpret. An example might be mapping nuclear radiation data from a meltdown with blues and greens (see Figure 5.5). When you say "radiation" to most people, they think of warmth—often extreme, very uncomfortable warmth. Thus, using a warm color scheme on such a map would probably be more appropriate. (Unless, of course, you are trying to downplay the leak to keep people calm. Then, blue and white might be a good choice.)

Primary Colors

The primary colors are red, green, and blue (see Figure 5.6). From these three, other colors can be created. Primary colors are also extremely bright. Children love primary colors, so if you are designing a Web map for a children's Web site, you should think about using a lot of primary colors. Pop art maps for the Web have been proposed by several French cartographers as well for those who are looking for a little more vibrancy in their maps (Christophe, Hoarau, Kasbarian, & Audusseau, 2012). Avoid combining red and green on the same map, as many people who are colorblind cannot tell the difference between the two.

Secondary Colors

Secondary colors represent combinations of the colors cyan, magenta, and yellow, including changes in their color value and saturation levels (see Figure 5.6).

Tertiary Colors

The colors that fall between the primary and secondary colors on the color wheel are tertiary colors (see Figure 5.6).

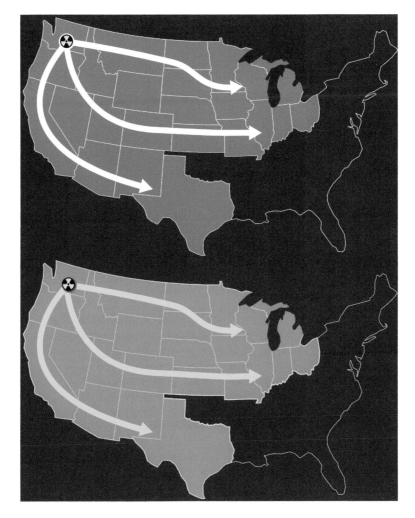

FIGURE 5.5
Fake nuclear disaster data represented with a cool color scheme (top map) and a warm color scheme (bottom map).

Monochromatic Colors

Monochromatic colors are created by manipulating the value or saturation of a single hue. Monochromatic colors are excellent for showing thematic data as lighter values intuitively indicate less of something (Figure 5.7).

Achromatic Colors

Achromatic colors are grayscale colors. They use only black, white, and gray values in between (see Figure 5.7). Although we often think of the Web as a

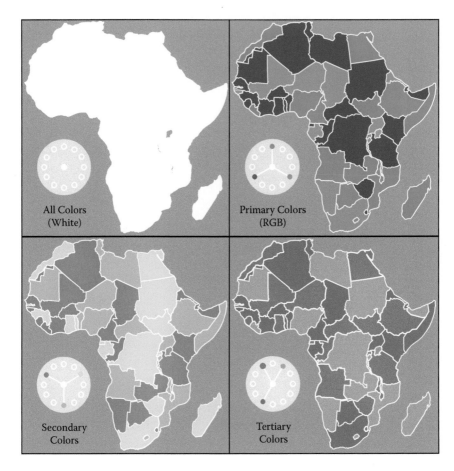

FIGURE 5.6

These four maps of Africa demonstrate several different types of harmonious color combinations that exist. The white map highlights that on screens the color white is created using all colors at full intensity.

cornucopia of color options, achromatic color schemes can be quite powerful and dramatic on Web maps as users expect lots of color and do not find it.

Complementary Colors

Complementary color schemes are those that make use of any two colors that are directly across from one another on the color wheel (see Figure 5.7). This includes these colors' tints and shades (i.e., manipulations of their values). Importantly, when complementary colors are placed next to or on top of one another, both of their saturations naturally increase. They will often start to visually vibrate. This leads to simultaneous contrast, a problematic phenomenon, which is discussed further in this chapter.

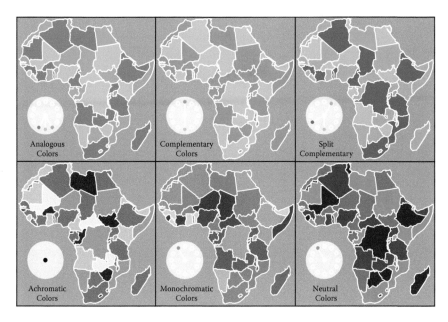

FIGURE 5.7

Six maps demonstrating different harmonious color relationships.

Split Complementary Colors

Split complementary color schemes are those that make use of a hue and any of the two colors adjacent to the hue's complementary color (see Figure 5.7).

Analogous

Analogous colors are those that are found next to one another on the color wheel. These can be used to show subtle differences. This is particularly effective for styling numerous features on a base map that you want to have present but do not want to emphasize at all in the visual hierarchy of the map (see Figure 5.7).

Neutral

Neutral color schemes are those that take a hue and create new colors by combining the original hue with the color black or its complementary color (Figure 5.7). This always results in darker colors. This scheme is great for designing very dark base maps or backgrounds.

Incongruous

Incongruous color schemes are those for which hue is chosen and then another nonadjacent hue, either to the right or the left of the complementary color, is also used. There are many interesting combinations that can be created by experimenting with incongruous colors.

Simultaneous Contrast: It Happens

Simultaneous contrast is something you need to be aware of when designing your maps. Simultaneous contrast is when two or more objects of the same color appear to have different values due to the colors surrounding them. Simultaneous contrast is particularly common when using aerial photos with lots of color variations as your base map. If some of the aerial imagery is lighter than other parts due to the nature of the landscape, thematic data or point symbols may appear to be a different value than the same symbols placed over another part of the map.

Color Rules That Remain Unchanged from Print

Certainly, color representation is very different between print and screen-based maps. However, many of the truisms you may have learned about color remain relevant today.

Less Color Variation Is More Powerful

It has long been known that you should limit the number of hues and values on your map to help users better interpret what they are viewing. This remains truer than ever with Web mapping, as in addition to your visualization, the map user is probably looking at colors in the browser's chrome, on his or her operating system (OS) desktop, and perhaps, on the rest of the Web page within which your map is embedded.

Therefore, it becomes imperative that in all but a few limited circumstances, you use neutral colors, or even gray scale, on your base map so that any colors you do use to symbolize important data catch the map user's attention. Some of the best colors to use for backgrounds are white, light gray, dark gray, or black. The use of too many highly saturated colors, particularly greens, in Web maps is the equivalent of cartographic pepper spray. It will likely drive map users away and leave a bad taste in their mouths.

Different Colors Have Different Meanings to Different People

Something that cannot be emphasized enough is that colors have emotional impacts on people. We have already mentioned how colors invoke temperature. Colors are also associated with different feelings, rituals, traditions, or experiences. Colors have an impact on how we think about things.

Colors and Feelings

The use of certain colors will have a major impact on how people interpret or receive your map's message. You should always choose a color scheme based on the intended audience. Think about which colors they are most likely to be receptive to. Are there colors that will help you tug on an emotional cord with the audience? (Orange maps for the Dutch, anyone?) These are questions that are not considered enough these days, particularly as it has become easy to choose a color scheme from an online color database. Some of these databases, such as Color Brewer (http://www.colorbrewer2.org), offer scientifically demonstrated, perceptibly different colors for clear and accurate communication. This makes it appealing to just borrow a scientific color palette and make a map.

I encourage you to create or adopt colors that will reinforce the message you are trying to communicate. Whether or not science has determined that one green value is sufficiently different from another so that your thematic map is accurately interpreted should only be *one aspect* of your color selection. Another, and perhaps more important one, is whether your color will subtly guide your audience to react to your message as you hope it will. For example, one should not use green when mapping casualties in the latest Middle East flare-up. Green in the west symbolizes peace and prosperity. Red would be a better choice. Red symbolizes anger, violence, and blood. Before you choose colors for your map, always ask yourself whether they complement the style and message of the map.

The Meaning of Color Varies by Culture

That the meaning of color varies by culture is particularly important to remember when designing maps for the Web because, typically, you will have no control over how far and wide your map spreads. Blue means water on many maps. Green means land. Travel to the Middle East, though, and Israeli maps often show Israeli territory as blue and Palestinian and Arab territory green. Thus, mapping casualties in the Middle East in green, as discussed in the previous paragraph, may imply that only one group of people is being killed. The most clichéd example is the color white. In Western societies, white is typically associated with purity, weddings, cleanliness, and all-around happy thoughts. In many Asian cultures, it is the color of death.

Just realize that certain colors will result in different interpretations depending on where and who views your map. Design for your intended audience. If you are designing for an audience in another country, be sure to look into the culture before you decide on a color. There are some great free resources online that you can visit to explore what colors typically mean to people in different places around the world. Some of these are shared in the "Further Reading and Resources" section at the end of this chapter.

Color Preferences Change over Time

The impact and meaning of color is time sensitive. Colors go through phases of popularity in different societies. In fact, you probably intuitively associate the 1960s, 1970s, 1980s, 1990s, and 2000s with different color palettes and styles. In Western societies, the color green was a very popular branding color through the first decade of the 2000s but is now rapidly disappearing, being replaced by blue (Markillie, 2011). It turns out that the green movement has saturated the market now (in the business, this has been referred to as "greenwashing"), and the color has lost its marketing luster. Instead, blue is the new green. Companies in the United States and Europe are increasingly using the color blue in their logos and media, as the color is associated with social collaboration and community cooperation. By the time this book is published, perhaps purple will be the new blue. The point is that one must keep up with broader color trends. Your map designs will likely benefit if your color selections are in tune with the trends of the day.

There is broad consensus in the world of Web design that the colors most used to exude a professional look on the Web are white, light gray or silver, black, and blue. This consensus is reified by the Web sites of most of the major tech companies, many of which have white-and-blue- or white-and-silver-themed Web sites. Thus, it stands to reason that these background colors on your maps, or ground colors if you will, will likely be perceived as attractive and professional to a large audience. As pertains to Web mapping specifically, white has become the background color of choice for many online Web maps and visualizations. For example, many of the *New York Times* maps and information graphics make use of lots of white space with subtle grays. Bright colors are used in contrast to the white backgrounds to highlight important data or data with which the user is currently interacting.

Color Models for the Web

The first step one must make when deciding on colors for any map is to choose an appropriate color model. Color models are systems for specifically defining different colors so that they are replicable across projects. There are numerous color models out there, several of which are described next. However, what needs to be noted right away is that you really only have one color model choice when it comes to Web mapping: RGB.

RGB (Red, Green, and Blue)

Any time you look at a screen, you are viewing the RGB color model. This is an additive color model. Red, green, and blue are combined at full intensity to create the color white. In fact, this is why staring at white screens for hours can be very hard on the eyes. Unlike with paper, which simply reflects the light present in a room, when viewing a backlit screen your eyes are being bombarded with the full intensity of three colors mixed to create white. The brightness is so intense that it can even cause insomnia and alter sleep patterns (Sutherland, 2013).

There are many different ways of expressing RGB colors in your Web maps. However, the most common methods are to use RGB decimal or RGB hexa-decimal (or #RRGGBB) model numbers. Regardless of which method you use to select your colors, you have up to 16,777,216 colors at your disposal.

RGB Decimal

RGB colors can be created using numbers. Each color—red, green, and blue—is given a value from 0 to 255. The lower the number, the less intense the color is; zero means the color is absent altogether. The higher a number value, the more intense the color is. All three colors combined at maximum intensity ($r = 255$, $g = 255$, and $b = 255$) equal white.

In addition, CSS3 (Cascading Style Sheets version 3) has added an alpha channel to RGB, making it RGBA. Thus, when using CSS to design your Web site and Web maps, you can also manipulate the transparency of RGB colors. Alpha is based on a 0–100 scale and, in most scripting languages, including CSS, is typically represented as a value between 0.0 and 1.0.

RGB Hexadecimal

RGB hexadecimal is probably the most common method of representing color in Web design. RGB hexadecimal breaks RGB decimal values down into six-digit, three-byte number codes that can be used with HTML, CSS, SVG (Scalable Vector Graphics), and many other applications. The format is more compact than RGB decimals and can be used to display just as many colors. (For example, the color white in decimal form is 255,255,255; in hexadecimal form, it is simply FFFFFF.) Luckily, there are hundreds of programs and Web sites that provide hexadecimal values for different colors; some are mentioned at the end of this chapter. You can typically pick a color and copy and paste the value into your script or the design program you are using.

The key is to understand how to read RGB hexadecimals so you can determine which approximate color a value will represent. They are six-digit, alphanumeric representations of all the 16,777,216 RGB colors at your disposal. Hexadecimal notation begins with a pound sign (#). This is followed by two hexadecimals representing the red value, followed by two more representing the green value and two more representing the blue value.

But, what is a hexadecimal? It is a 16-bit representation of a number, as opposed to the more standard 10-bit decimal representation that we use

Hexadecimal = 1,2,3,4,5,6,7,8,9,A,B,C,D,E,F
Decimal = 1,2,3,4,5,6,7,8,9,10,11,12,13,14,15

Tomato Red = R255 G99 B71

255/16 = F (numeric 15)
    Remainder = F (numeric 15)
    R = FF

99/16 = 6
    Remainder = 3
    G = 63

71/16 = 4
    Remainder = 7
    B = 47

Tomato Red = # FF 63 47

FIGURE 5.8

An example of taking an RGB decimal value for a tomato red color and transforming it into an RGB hexadecimal value. Generally, it is faster to use a free online resource (listed at the end of the chapter); however, anyone can do it in a pinch.

on a daily basis. In hexadecimal notation, the numbers 0–9 are represented using the numerals 0–9. However, the numbers 10–15 are represented with the letters A–F. One can calculate the hexadecimal notation of a decimal RGB color simply by dividing the decimal values by 16 (see Figure 5.8).

Hue, Saturation, and Lightness

HSL (hue, saturation, and lightness) is an additional RGB color space that is popular among computer scientists because it is arguably more intuitive to work with than RGB. It is not to be confused with HSV (hue, saturation, and value), which is similar in how it organizes colors but can vary significantly from HSL. HSL is now included as a color specification in CSS3 and thus can easily be incorporated into Web maps. However, its compatibility across browsers and devices is less reliable than RGB because HSL was only just added as a styling option. As with RGB in CSS3, HSL is also allowed to have an alpha channel (becoming HSLA), so one can use the color model to manipulate a color's transparency.

RGB Colors Will Look Different Depending on the Screen

Unfortunately, backlit monitors will typically display RGB colors differently depending on the screen technology being used. Thus, it is important to note that no matter how well you plan the colors of your Web map, you do not have absolute control over how your map will look to end users.

Never Design Web Maps with CMYK

The CMYK (cyan, magenta, yellow, and black) color model is one of the most commonly used in the world today. Nearly everything printed on a paper medium uses this model. It is a subtractive color model, which means color is created by subtracting from white. As one adds ink to a piece of paper, less light is reflected, and different colors appear. This color model mixes three colors (cyan, magenta, and yellow, or CMY) to create all of the colors you see in print. Although all colors can be created with just CMY, black (or K) is added to create more contrast.

You should never use the CMYK model for Web maps. Here is why: CMYK cannot be accurately represented on a screen device. Monitors do not subtract to make color; they add. Every screen you look at, from your mobile phone to your high-definition television, uses the RGB model to represent color. (Okay, yes, there are a handful of televisions by Sharp that add a fourth color, yellow.) CMYK colors presented on screens will look different than they do in print. Therefore, designing your Web map with CMYK will likely result in undesired color appearance.

Coloring Your Base Map

Once you have chosen your color model and determined what colors are likely to appeal to your desired audience, it is time to start creating your colors. The rest of this chapter concentrates on helping you achieve this goal with some tips for selecting effective colors for your base map, reference map, and thematic mapping needs.

Solid-Color Base Maps

As mentioned, certain background colors are generally more effective than others. White and light gray are great choices for base map colors because they act as a neutral canvas for other map symbols to jump out as figures. Lighter backgrounds help colors on your map stand out—as long as you are not using too many colors—and often make it easier to read text. You can rarely go wrong using light gray as your base map color (see Figure 5.9).

Again, audience should guide one's decision process about background color. If your map is being designed for a group of people who have only used traditional paper maps or are not likely to be connoisseurs of Web maps, a simple, light-colored background is a smart choice. On the other hand, if your map

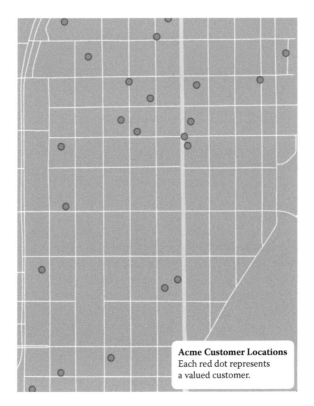

Acme Customer Locations
Each red dot represents
a valued customer.

FIGURE 5.9
Gray map backgrounds allow colors in the foreground to pop out.

needs to have some "wow" factor, you may want to unsettle map users' expectations by inversing the figure-ground color scheme. You can do this by making the base map and map background dark and using light colors to make features stand out (see Figure 5.10). Dark backgrounds are particularly effective for emphasizing a handful of symbols on a map. Just remember, if you are not designing a reference map, the elements of a base map should fall very low in the visual hierarchy. Always avoid using too many colors in your base maps.

Shaded Relief

Of course, what if you want to include shaded relief? There are generally two ways to do this: grayscale shaded relief or hypsometric tinting. Hypsometric tinting shows elevation change through the use of a color scheme. Depending on the topography of the area you are mapping and the gradient scheme you choose, hypsometric tinting can run the gamut of colors. Figure 5.11 illustrates the visual difference between a fully colored hypsometric relief representation, the same relief with 33% and 66% transparency, and the same

Acme Customer Locations
Each red dot represents
a valued customer.

FIGURE 5.10
Dark map backgrounds used in conjunction with bright, contrasting colors can be effective for creating dynamism.

relief in gray scale. Hypsometric tinting can be wonderful, particularly if it is highlighting an important component of your map's message and is done using subdued colors. However, it can also add additional cognitive strain on a map user and make it difficult to symbolize important information. Often, map users will misinterpret hypsometric tinting for ground cover. Unless topography is a key aspect of your communication goals, always deemphasize it or simply eliminate it.

Aerial Photography

Web maps without dark-tinted aerial photography are becoming increasingly rare. This is not necessarily a good thing. The traditional goal of a cartographer has been to abstract reality to make it less complex and easier for map users to make sense of their environment. Some have argued that interactive, Web, and multimedia mapping has turned this original goal of

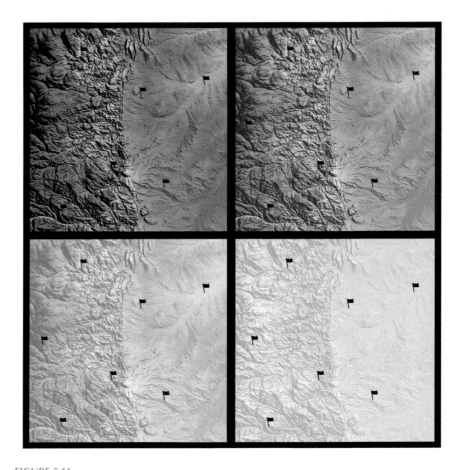

FIGURE 5.11

Hypsometric tints are fancy but often too bright for map backgrounds in Web maps. Gray-scale shaded relief often works much better. On which map can you most quickly assess the locations of all of the flag icons? Clockwise from top left: fully-saturated hypsometric tint; 66% opacity hypsometric tint; gray-scale shaded relief from the same image file; and 33% opacity hypsometric tint.

mapping on its head. For many GIS (geographic information system) and map service experts, the goal now is to create a map that mimics reality as closely as possible so people can make extremely informed decisions about their environments based on real-time data. As such, tiles of aerial photography and satellite imagery with streets and certain landmarks highlighted are increasingly becoming the default view in Web maps. It is believed that people will quickly be able to recognize landmarks from satellite imagery and have a better grasp of locations and spatial relationships in a given mapped environment when making decisions. Some companies are even taking 360-degree lidar data of cities to add to their maps (Kelion, 2012). Others, however, most notably neuroscientists and researchers who study human cognition, note that more often than not having too much information at our disposal leads to far poorer decision making than if we only have a limited amount of data (Gigerenzer, 2007; Gladwell, 2005). Anyone who has walked past the coffee shop they are looking for while furiously attempting to follow and figure out the location of a blue dot on their mobile phone map can relate.

Aerial photography can be great in certain circumstances (i.e., if you want to spy on what your neighbors have been building in their backyards). However, for most mapping endeavors, such photography simply distracts from the data and information you are trying to present. It is a gratuitous display of data that does not help one's communicative goals.

Regardless of how neat it looks, using colored aerial imagery as part of your base map will also make it difficult to make the important information more noticeable. If you have to use aerial imagery, tone down its alpha values or mask it with white so that your symbols and data stand out. If such imagery does not help you achieve your communication goals but your client insists you include it, then a smart workaround is to set the imagery as a layer option. Let your map users decide if they *want* to turn it on. Figure 5.12 demonstrates how distracting aerial photography can be if it is not at least masked somewhat or left out altogether.

Reference Map Color Schemes

Reference map color schemes have truly begun to evolve from print cartography. Color schemes have always varied by culture and society. For example, autobahns on some maps in Germany are represented with blue, whereas in the United States, most students are taught that blue lines should never be used for anything but rivers. However, today we are starting to see a general universalization of certain reference map color schemes. This is mostly due to the fact that the largest mapping services in the world, owned by some of the world's largest tech companies, have standardized their own maps and have settled on similar color palettes. Although each service differs slightly in its color choices, a consensus has emerged. For example, major roads are typically represented with a yellow or purple color value; minor roads are typically a shade of black. Figure 5.13 highlights the Google Maps street theme.

Since many of these map service provider color choices have come to be expected by online map users, generally, it is a good idea to use these or a similar color scheme on your reference maps. Maps are always easier to read when users do not have to spend time learning a new set of symbology and color references. Familiarity helps map users pick up on visual cues of what they are viewing.

FIGURE 5.12
Aerial photography is cool to look at, but not always very useful to humans who rarely look at the world from straight above. The best maps abstract information to make it easier to understand. Aerial photos add complexity and, most importantly here, make it very difficult for symbols and colors to stand out. The map on the left is a Google map of several places of interest in Pecs, Hungary. The map on the right is the same map with Google's satellite imagery. Map© 2013 Google. Imagery© 2013 DigitalGlobe.

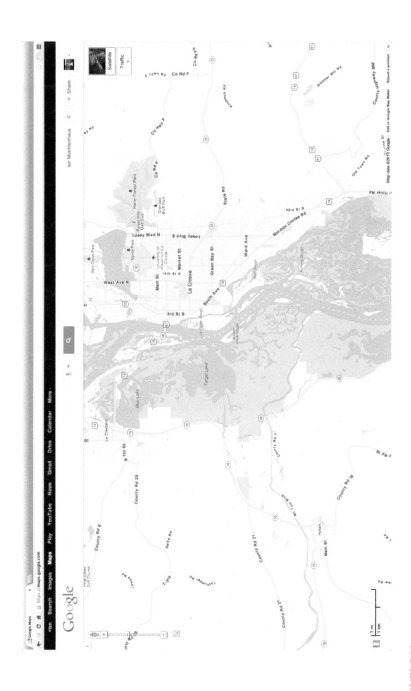

FIGURE 5.13

A screen capture of Google Maps, taken on March 19, 2013, of La Crosse, Wisconsin, USA.© Google.

When to Ignore Color Conventions

There certainly are some benefits to creating your own, novel color schemes. In fact, some map service providers' color schemes are not that aesthetically pleasing and not very stylish. Thus, experimenting with reference map color is encouraged. However, if your newly created colors are not in line with your communication goals, map purpose, or increasing your map's rhetorical appeal, then stick with the standard color scheme for your reference map.

Novelty Equals Attention

One reason to create your reference map color schemes from scratch is to attract attention to your map. Nothing stands out like novelty. Bold and unusual color choices on a reference map will often excite a map user. Surprise works well to attract someone's attention.

Does the Color Scheme Match the Message?

Another mapping issue that tends to be overlooked these days, perhaps because of the ubiquity of APIs (application programming interface) with built-in, premade reference maps, is whether a map's color scheme works well within the Web site, mobile app, or operating system in which it is being displayed. It is good practice to style your reference map to match the encompassing medium. If your map is going to be embedded in an iOS app, for example, use colors that work well and complement the graphical user interface (GUI) of the app itself. If your map is going to be embedded in a nonprofit Web site, use reference map colors that are the same hue or of similar color value as the Web site. Your map communication will be more effective if your design fits within the cohesive whole of its packaging (e.g., the Web site or your mobile app).

Resources for Finding Colors

Many APIs allow you to change the look of your base map using JavaScript or other scripting languages. CloudMade, the provider of the Leaflet API discussed in Chapter 12, has created an interactive, intuitive base map color editor. Using this online tool, you can style freely available OpenStreetMap data however you like and export the styles for use in your maps using the Leaflet API. No scripting is required; CloudMade does it for you. Figure 5.14 shows some of the numerous styles you can use or adapt from right on the CloudMade Map Style Editor Web site.

Another great resource for finding colors that work well together aesthetically is Adobe Kuler (https://kuler.adobe.com). People from around the world have created color palettes based on their favorite artwork or

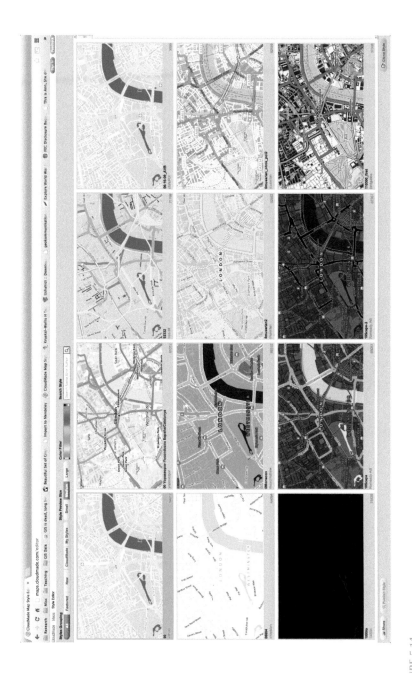

FIGURE 5.14

A screen capture of CloudMade Map Style Editor Website taken on March 19, 2013.© CloudMade (www.cloudmade.com).

paintings or just by playing around and have shared them on this site. Here, you can find some color palettes you like, adapt the colors on the Kuler Web site to better suit your map needs, and export the colors to Adobe programs or copy and paste the RGB codes right into your HTML, CSS, SVG, or other file (see Figure 5.15).

Choosing Thematic Map Colors

Selecting effective thematic map colors is a different process entirely from selecting reference map colors. If you must design clearly distinct colors to effectively communicate your data, your first stop might be Color Brewer (http://www.colorbrewer2.org). This site allows you to experiment with previously derived color schemes. The schemes are comprised of perceptibly different colors so that people can view your data clearly. It is a great resource for mapping data that rely heavily on color variation. Of particular use when it comes to Web mapping is the fact that you can filter color schemes so that only those that work well on screens (i.e., via RGB) are available for selection (see Figure 5.16).

The point of Color Brewer is not to design necessarily aesthetically pleasing palettes but distinct and clear palettes. However, if the thematic data do not necessarily need to be read for clarity but are included in a map to achieve a rhetorical goal or purpose, designing your own color scheme to better invoke a message is extremely advisable (e.g., toxic greens for potential nuclear fallout). There are myriad tools available, such as the two mentioned previously, that allow you to experiment with colors and copy–paste or export the RGB codes for use in your map. Links to these and additional resources are included at the end of this chapter.

Beware of Color Blindness

One thing that needs to be mentioned is that certain colors should simply never be used together. Approximately 10–14% of all males are red–green colorblind. In addition, there are several other types of colorblindness, including people who cannot discern blue from yellow. You should never, *ever*, design maps that use different reds and greens together to symbolize different objects or data values (see Figure 5.17). About 5–10% of your map users will not be able to make sense of your map if you do.

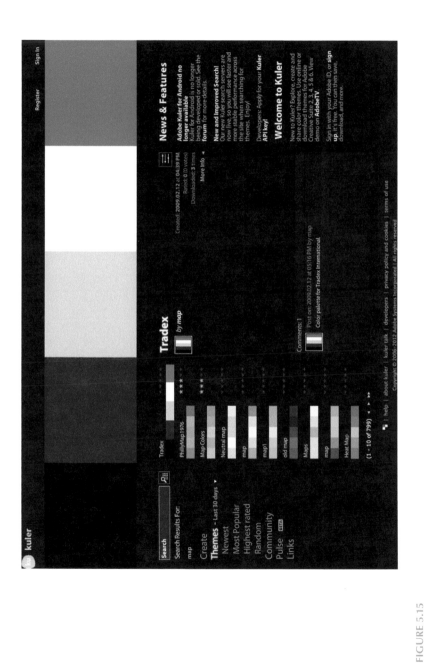

FIGURE 5.15

A screen capture of Adobe Kuler Web site taken on March 19, 2013. (Used with permission.© Adobe, Inc.)

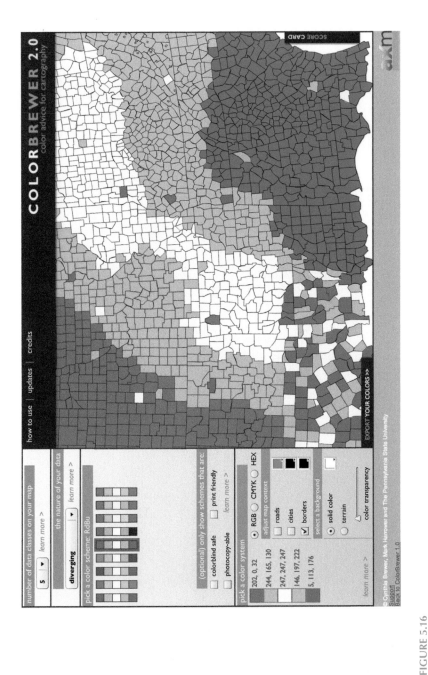

FIGURE 5.16

A screen capture of the Color Brewer Web site (second edition) taken on March 19, 2013. (Used with permission. Created by Cynthia Brewer and Axis Maps.)

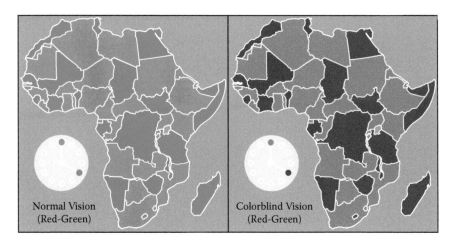

FIGURE 5.17

These two maps show exactly the same information using exactly the same colors. The map on the right is how this map appears to people with deuteranopia (red-green color blindness). See the reference to VisCheck at the end of this chapter to colorblind test your own maps.

Key Concepts

- Color remains one of the most important design elements in map-making; it is also the least stable. Colors will appear different for a variety of reasons, including the technology being used and the lighting in a room.
- People perceive colors differently.
- Color schemes are trendy; colors that were exceedingly fashionable and professional looking last year may be antiquarian by next.
- Choose your color scheme first and foremost on how well it will help you achieve your communicative goals.
- Do not just reinvent color schemes for no reason. Certain color conventions are increasingly being expected by Web map users (e.g., some roads in yellow). Never go against normal conventions just because you can.
- Thinking about human perception is key to the effective use of color on maps. Be aware of accidental simultaneous contrast. Remember never to use red and green together. Recall that less color contrast among features on a base map gives you the power to use color to make important data stand out.
- Neutral or dark base maps help colors pop out on your map.
- Avoid the use of satellite imagery as a background.

- Many people are colorblind (i.e., up to 12% of males). *Never use red and green* of the same saturation and value to represent different types of data on your maps.

Further Reading and Resources

Online Color Resources

Adobe Kuler. https://kuler.adobe.com/. This Web site is a great place to go for color inspiration or to create your own color palettes. You can create patterns and swatch palettes based on your favorite art and pictures here. You can also download thousands of premade and shared swatch files and RGB values right into your project. If you use Adobe software, Adobe Kuler applications are built right into many of the design programs (e.g., Adobe Illustrator and Adobe Photoshop).

Color Brewer. http://www.colorbrewer2.com. Color Brewer is an extremely useful resource when designing color palettes for thematic mapping. Some of the color palettes are more attractive than others, but all have been tested to help make sure that differences in data values are perceptible to most map users. You can download the color palettes into a variety of formats and copy and paste RGB values right from the site.

Vischeck. http://www.vischeck.com. Vischeck is a free Web site and a great place to visit if you are wondering how your map might look to someone who is colorblind. You simply type a Web site address into a dialog, and it shows you how it looks to someone with one of three kinds of color blindness. (You can also catch a glimpse of how babies see your map, which makes for a fun bit of procrastination.) It is hoped the results reinforce the fact that you should never use red and green together on the same map to differentiate information. Try it.

Color Meanings and Emotions

Cousins, C. (2012, April 3). Color and emotions: what does each hue mean? http://tympanus.net/codrops/2012/04/03/color-and-emotion-what-does-each-hue-mean/ (short URL: http://goo.gl/P65lS).

QSX Software Group. Color Wheel Pro–see color theory in action. http://www.color-wheel-pro.com/color-meaning.html (short URL: http://goo.gl/xZAc1).

Color Pickers

Colorpicker.com. Home page. http://www.colorpicker.com/.

W3Schools. HTML color picker. http://www.w3schools.com/tags/ref_colorpicker.asp (short URL: http://goo.gl/w6c0a).

Color Style Tools for Different Map Services

CloudMade. OpenStreetMap Style Editor. http://maps.cloudmade.com/editor.
Google Maps API Styled Maps Wizard. How to use the Styled Maps Wizard. http://gmaps-samples-v3.googlecode.com/s vn/trunk/styledmaps/wiza rd/index.html (short URL: http://goo.gl/rmRxj).

Hexadecimal Color Creator

Colorspire. (2013, March 27). RGB color wheel. http://www.colorspire.com/rgb-color-wheel/.
Irby, L. (2013). Hex color code chart & generator. http://www.2createawebsite. com/build/hex-colors.html#colorscheme (short URL: http://goo.gl/V0b9H).

Further Reading

Carter, R. (2002). *Digital color and type*. Crans-Pres-Celigny, Switzerland: RotoVision.
Christophe, S., Hoarau, C., Kasbarian, A., & Audusseau, A. (2012). A framework for pop art maps design. GIScience Conference (p. 4). Columbus, OH. Retrieved from http://www.giscience.org/proceedings/abstracts/giscience2012_paper_83.pdf.
Gigerenzer, G. (2007). *Gut feelings: the intelligence of the unconscious* (p. 280). New York: Penguin Group.
Gladwell, M. (2005). *Blink: the power of thinking without thinking* (p. 296). New York: Back Bay Books.
Kelion, L. (2012). Here's how Nokia creates its maps. *BBC News*. Retrieved March 19, 2013, from http://www.bbc.co.uk /news/technology-20497719.
Markillie, P. (2011). The greening of blue. *The Economist*. Retrieved from http://www.economist.com/node/21537977.
Sutherland, S. (2013, February 1). Bright screens could delay bedtime. *Scientific American Mind*, 13.

6

Typography

Introduction

This chapter is concerned with separating which typography rules from print maps have largely stayed the same in Web maps from those that have changed. First, we review those type rules that have not changed, before covering all that have. Then, we conclude with a brief overview of several fonts that are ideally suited for Web mapping.

All in the Family: Explaining Typeface and Font

Officially, a typeface is not a font, and a font is not a typeface. However, the word *font* has largely come to replace the term *typeface* among the general public. Officially, a typeface is only one style of type. For example, Myriad Pro Regular is a different typeface from Myriad Pro Semibold. Both of these typefaces belong to the same typeface family (i.e., Myriad Pro), but they are distinct. A font, on the other hand, officially refers to typeface of a particular size. One way to think of it is as follows: If a typeface is a pair of shoes, a font represents those shoes in a particular size.

Modern computing has certainly not helped stymie the inappropriate use of the word font. In fact, most operating systems have "font"—not "typeface"—folders. People code HTML and CSS (Cascading Style Sheets) using the word *font*, not typeface. Finally, almost every single word-processing program has you select a font when you want to type something. While some type connoisseurs get persnickety about the inappropriate use of the word, I am going to use the word *font* in this manuscript to stay in line with what is used in other software programs. Also, I use *font styles* to refer to specific typefaces within a typeface family (e.g., Corbel bold, Corbel italic, Corbel regular). Last, I use *font size* to refer to a typeface's size.

Typographic Rules That Stand Firm

The Web has revolutionized how type can be used on maps. No longer does text need to be static, but it can be resized, manipulated, and even created by map users. Although this has meant that many things about map typography have changed, most of the print standards dealing with labeling and text placement stay true today.

No More Than Two Fonts per Map

Sure, there are always exceptions to the rule that there should be no more than two fonts per map. Sometimes, you may want to use more fonts to better approximate a vintage-looking map. Occasionally, you have a client who wants you to use five distinct fonts. (Never forget who pays the bills.) A good rule of thumb, however, is never to use more than two fonts on a map. In many cases, though, you should be able to get away with using a single font. All other qualitative or quantitative differences expressed by the fonts you choose should be made by manipulating a font style (e.g., bold, italic) or properties (e.g., font size, color, capitalization).

Font Styles and Properties Indicate Different Things

Differences in type size, case, and style still have the same visual impact on the Web map user as they do in print maps. Certain attributes should only be used to represent qualitative differences between what is being labeled (e.g., text color). Other manipulations, such as font size, can be used to represent quantitative distinctions, such as bigger cities compared to smaller ones. See Figure 6.1 for a quick overview of the variations in type used to represent different data attributes.

Text as a Core Map Element

Many cartographers describe map labels and supplemental information boxes as core map elements within a map's visual hierarchy. This remains true for Web maps. Text and labeling—both its design and its placement—are key design elements in every form of geocommunication. Many cartographic standards regarding type have been forged over the past several hundred years and should not be ignored.

Labeling Norms

If you are coming from a print map background, Figure 6.2 is probably common knowledge; it offers a quick primer on the key rules for labeling different

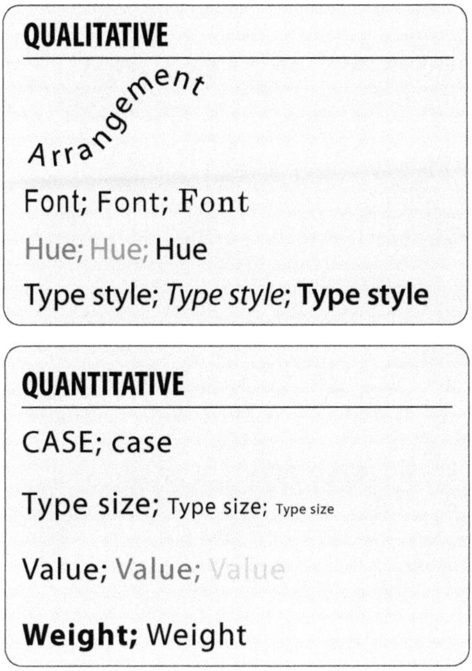

FIGURE 6.1

Common visual variables showing qualitative or quantitative differences with type.

spatial features. I highly recommend that you memorize the ideal placement of point labels. Also, remember to label linear features on smooth stretches. For example, do not label a river where it winds and bends a lot. Offset your label from the linear feature a little bit to give the text some breathing space. Finally, for areal (i.e., large polygonal) features, it is often a good idea to use letter spacing. Web maps that incorporate these ideals will be easier to read and more soothing to most map users in Western societies.

Typographic Rules That Have Changed

Although type styling may be largely missing in Web maps, many changes in Web typography have been for the better. As recently as 10 years ago, making use of decent fonts in HTML was hit or miss. In fact, this was one reason that Macromedia Flash, now owned by Adobe, took off as an interactive Web plug-in. With Flash, people were able not only to create quick animations but also to embed whichever fonts they desired in their Web and map design. Today, with the full development of HTML5 on the horizon and

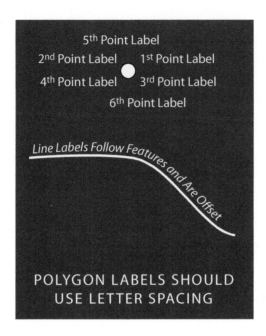

FIGURE 6.2
A brief overview of the rules of point, line, and polygon labeling. (For a more detailed overview of print map labeling, refer to *Thematic Cartography and Geovisualization* by Slocum et al., 2008.)

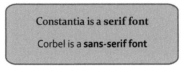

FIGURE 6.3
Serif versus sans-serif font.

the power of CSS, we can do far more with fonts. Next, I review several issues you need to be aware of when you start planning the type styles and fonts you will use in your Web map.

When Possible, Ditch Serif Fonts

There is a new cartographic convention that has evolved with Web cartography: *use only sans-serif fonts* (Figure 6.3). The avoidance of fonts with serifs in Web maps developed with good reason; resolution on most digital media is mediocre. Serif fonts are actually more difficult to read on backlit displays (i.e., LCDs, cathode-ray tubes [CRTs]). Even as recently as 2005, a typical resolution for digital displays was 72 ppi (pixels per inch). With so few pixels, type serifs often ended up looking pixelated or disappeared just enough to make fonts less than legible. Additional research has indicated that serif fonts on digital devices blur spacing between letters (Buckley, 2012). Finally, there is a cultural reason. In Western societies, sans-serif fonts are typically considered more contemporary and professional looking than serif fonts. Thus, sans-serif fonts are going to be employed more by those designing modern-looking Web maps.

Within 10 years, most of this concern about serifs and Web maps will probably be irrelevant due to the advancement of high-resolution smartphones, tablets, and laptops. Nevertheless, the standard has already been set: Sans-serif fonts are the norm for Web mapping.

Size Does Matter

Not all fonts are created equally. The best fonts for screens have generous x-heights, spacing between letters, internal letter widths, and consistent thicknesses (see Figure 6.4). The x-height refers to the height of a lowercase *x* character in the font. The higher the x-height, the easier a font is to read on the screen. Spacing between letters refers to the amount of white space between individual letters. Reliable Web fonts have larger spaces between each letter, which also makes them easier to read. The wider spacing also helps ameliorate issues arising from low-resolution displays that can make individual letters more difficult to discern. Internal letter width, or punch width, represents the thickness of spaces within the letters (see Figure 6.4). The wider the punch width, the more pronounced and distinct a letter will appear on a monitor. Finally, consistent thickness also helps make a Web font more legible. Consistent line weight and thickness, with simple strokes, help make text stand out clearly. (In fact, varying font thickness is one reason many serif fonts do not look as smooth on screens.)

Font Size Is Now Voodoo Science

Gone are the days of simply stating that no map should have a font under 6 points. High-resolution devices are making a mess of determining how

FIGURE 6.4
The anatomy of text.

large your maps will appear on digital devices, which also has a direct impact on the sizing of your fonts. This is a real problem with no end in sight. One nice thing about standard computer displays only showing 72 or 96 ppi was that Web developers generally had a good sense of how large things would look on different devices. Enter the era of mobile devices (i.e., smartphones and tablets) with increasing resolutions, and suddenly a font that appeared large enough on your home computer screen is now being read on your friend's 4-cm smartphone screen.

Font size can vary dramatically on devices with different resolutions and pixels per inch (see Figure 6.5). What makes things even more complicated

FIGURE 6.5
Example of how much text can shrink between devices with low and high pixels per inch.

is devices are coming out with varying resolutions nearly every month. One workaround is to add the ability for a user to manipulate font sizes on a map. Doing this is not always a simple endeavor, particularly when it comes to maintaining the integrity of your labels, but it is almost always beneficial. The alternative is to design your fonts with high pixels-per-inch devices in mind from the beginning, ensuring that the font can only increase in size (i.e., when it is viewed on devices with fewer pixels per inch).

Not Everyone Has Access to the Fonts You Do

Web mapping without a plug-in (e.g., Adobe Flash, Microsoft Silverlight, Oracle Java) depends on HTML, CSS, and scripting languages (i.e., primarily JavaScript). HTML and CSS make use of fonts found on your map user's computer. Different computers and mobile devices will have different fonts installed on them. A prime example of this can be found in the common fonts found on Apple and Windows personal computers (PCs). Helvetica, a proprietary font, is frequently found on Apple computers; it is missing from almost all Windows computers. Many Linux operating systems are even more limited in their proprietary font choices.

  If you design a map that loads labels separately from a tiled background, you will want to make sure that you do not use only a single obscure or specialized font in your CSS. If your map user does not have the desired font on his or her device, a default font will be chosen and might therefore be less aesthetically pleasing. In CSS, it is fairly easy to provide an ordered list of which fonts are most desirable for your map or Web page (see Figure 6.6). Providing backup fonts, as is done in Figure 6.6, will provide you more control over what a map user sees. Unfortunately, you will never have absolute control over how fonts are displayed on end-users' computers.

Will Your Map Have Interactive Text?

One thing you never had to consider when dealing with text in print is whether you wanted it to be a graphic that is immune to user attempts to interact with it (i.e., image text), selectable (and therefore highlighted), or interactive (a button that does something, animated text, or a hyperlink). These are important things to keep in mind and plan for *before* you start creating your map.

```
body {
    font-family: Calibri, Corbel, Verdana, Geneva, sans-serif;
}
```

FIGURE 6.6
CSS (Cascading Style Sheets) code to create an order of desired font choices.

Image text is text that is not at all responsive to user interaction. Most map labels are of the image text variety. Note that this does not mean that the text is necessarily embedded as a picture or in raster form—although it can be. It just means that the text is not selectable, and the text is part of the underlying map image.

Selectable text is text that a user can highlight with a mouse or touch device and then copy. Sometimes, it may be advantageous to make text selectable so that people can copy information they find useful and paste it into another application (e.g., an address). However, in certain circumstances, you may not want people to be able to select and copy the text displayed. Perhaps some of the text shows data that are proprietary and only to be viewed within the map. Or, maybe you are worried that making text selectable will become distracting. Often, maps use text that is selectable by default. Every time someone tries to pan a map by dragging the mouse, for example, different text will highlight, and the map will not pan. This can be confusing and irritating. Make sure you only make text selectable that will be of potential use to map users.

Interactive text is text that users can control, manipulate, or interact with in some fashion. Interactive text takes many forms. For example, perhaps you have text that acts like a button. When someone clicks with a mouse or taps with a finger on the text, something happens (e.g., information loads into an info window or a national anthem starts playing). Beyond acting like buttons, text is interactive if the map user can manipulate how it looks using a graphical user interface (GUI) to change the appearance of text (see Figure 6.7 for examples).

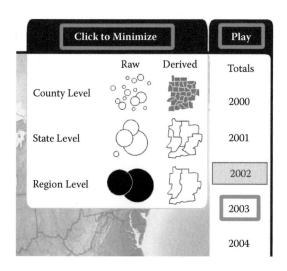

FIGURE 6.7

Example of interactive text (i.e., text that functions as a button). Several options for selecting items are highlighted in red. (Map courtesy of Dooley and Muehlenhaus, 2010.)

Fonts That Play Well with Web Maps

There are some classic Web fonts that you can assume a vast majority of your Web map users will be able to access. Some of these are discussed next.

Verdana

Verdana is one of the most popular sans-serif fonts used in Web design (Figure 6.8). It has a large x-height, wide punch widths, and adequate spacing. The line thickness varies a bit but not drastically. It is available on all Windows machines.

Century Gothic

Century Gothic is a professional-looking font. It has much thinner strokes than Verdana (Figure 6.8). Its spacing and punch width are ideally suited for clear reading. This is best for larger-size fonts.

Arial and Helvetica

Arial is one of the most used, and overused, fonts in the world. Helvetica is the closest version to Arial found on all Macintoshes. Although invented years before the rise of the modern Internet, both of these fonts have large x-heights, making them great for online displays (Figure 6.8). The only downside is that almost everyone uses these fonts in their Web pages. Thus, the fonts lack aesthetic uniqueness.

> Verdana
> Century Gothic
> Arial
> Helvetica
> Trebuchet MS
> Tahoma
> Corbel
> Myriad Pro
> Myriad Web
> Georgia
> Palatino

FIGURE 6.8
Examples of various fonts that work well on Web maps.

Trebuchet MS

The Trebuchet MS font is stellar in a variety of ways (Figure 6.8). It has distinct letter shapes (check out the capital *M* for example). It has a large x-height and great punch width, and it has clear letter spacing. This font is a little more unusual than other sans-serif fonts and may look odd on some maps. However, its uniqueness is also a potential strength when it comes to angling for an original look to your online maps.

Tahoma

Tahoma is often grouped with Verdana when it comes to the discussion of fonts (Figure 6.8). Both were released by Microsoft in the mid-1990s, and at first glance they appear fairly similar. There are some major differences, however. Tahoma has far tighter letter spacing than Verdana. The punch widths and letter widths are smaller as well. This font is more compact and feels more cramped than Verdana. However, this has its benefits. If you need to place many labels on your Web map, Tahoma will take up less space. It will be harder to read on a screen than Verdana, however.

Corbel

Corbel is another great font and a personal favorite of mine. It was developed by Microsoft and is on every post-Windows-XP PC and any computer that has Microsoft Office 2007 or later installed. It is also available for free via the PowerPoint 2007 viewer (which is how many Linux users get their hands on it). This font was developed for the screen and has some unique features compared to other sans-serif fonts. Like the others, it has decent punch width, efficient letter spacing, and adequate x-height. Most useful to cartographers, however, is that Corbel was designed to be legible at smaller font sizes on screens.

Another interesting feature of Corbel font, and the reason it is one of my favorite fonts, is that it uses lowercase numerals (also known as "old style"). Lowercase numerals are numbers that have varying heights; some of the numerals remain at the x-height; others ascend beyond the x-height, and others descend below the x-height (see Figure 6.8 for an example). Lowercase numerals are rare in contemporary fonts, particularly those designed for screens. There is a widely held belief that lowercase numerals are easier to read (particularly in block text). They add variation just like lowercase letters do. They can also give the impression of high-quality printing, as mostly fine newspapers, magazines, and expensive books have used these numerals. Insert fonts using lowercase numerals into your Web map or legend, and you may be surprised how cool they will look.

Myriad Pro and Myriad Web

Adobe's Myriad Pro has become ubiquitous and is probably a bit overused as a default font by designers (see Figure 6.8). Myriad is really versatile, distinguished, and easy to read on screen and paper. Myriad Web is a version of this font that was redesigned for the Web. The letters are slightly wider than Myriad Pro's, making it easier to read on screens.

If this is such a fashionable font, why is it not used everywhere on the Web? One reason is that many people may not have it installed on their computers. Myriad Web needs to be purchased from Adobe. (The Web version of the font tends to be installed whenever someone installs part of the Adobe Creative Suite package, but many people do not have that on their computers.) So, when using Myriad Pro or Myriad Web, it is advisable to embed the font via coding (e.g., using CSS3 "@font-face") or plug-in (e.g., embed the font within Adobe Flash). If neither of these options works, you can always use it on rasterized or tiled base maps.

Georgia

If you want to use a serif font on your Web map, Georgia is a good choice (see Figure 6.8). This font was developed by Microsoft in the mid-1990s as the serif counterpart to Verdana. It is similar to Times New Roman, a serif default in print for decades. However, it was designed as a Web font and therefore has some major enhancements. Compared to Times New Roman, Georgia has wider letters, more consistent line thickness, less-ornate serifs, and a much larger x-height. One other major difference between Georgia and Times New Roman is that, like Corbel, Georgia has lowercase numerals. Probably the biggest benefit of the Georgia font is that almost every computer in the world comes with this font, or a close variant, installed.

Palatino

Palatino is one of the most widely used fonts in the world (see Figure 6.8). It was arguably the most used word-processing serif font in the world until Times New Roman passed it in the 1990s. It has been widely distributed; every Windows machine since 2000 and Microsoft Office installation since 2003 have provided this font. The font has decent letter spacing, but its lines can be a bit difficult to read on screen when the font is small. It is a good alternative to Georgia if you want a serif font that is a little less dominant as it is comprised of thinner strokes.

Comparing Type on Your Web Map

A great site to test and compare what different fonts will look like on your Web map or Web site is Typetester (http://www.typetester.org). You can select fonts found on your computer and compare them side by side with other fonts. More important, you can change certain aspects of the font, including both text color and background color. Choosing an appropriate and readable text color is as instrumental as choosing an appropriate font if you want to make your map a successful communication tool.

Font Myths, Realities, and Web Maps

This chapter presented a handful of general rules, guidelines, and suggestions for using legible and attractive fonts in your Web maps. However, rules are often meant to be broken. It is always a good idea to play to map users' expectations and standard mapping conventions, but it is also important occasionally to question and push back against some of the norms.

Map purpose and the aesthetics of your map should drive your font and labeling decisions. For example, if you have been hired to design an interactive map showing the diffusion of Viking culture, you may very well want to use serif fonts that better tie the map to the historical period the map represents. Just make sure the fonts are legible and work with the theme being mapped. That is the most important thing.

Another common myth among cartographers these days is that you should avoid using all uppercase letters (all caps) whenever possible. While this is sage advice when you are writing large blocks of text, avoiding all caps at all costs is not a wise strategy when it comes to visual communication. All caps can be used successfully to help differentiate between features. Increasingly, the myth that all caps are more difficult to read than lowercase letters seems to be scaring cartographers and others away from using this font attribute at all. For example, on many prominent Web maps, the names of countries are in mixed case and often start to meld with city and province names. (The argument that capital letters are harder to read was first hypothesized in the 1880s by a man named James Cattell. Numerous studies have since discredited his hypothesis; all caps are not more difficult to read. We just read them more slowly. Check out the 2011 work of Weinschenk for more information on this text myth.)

Conclusion

Two things have not changed with the rise of Web mapping: Font selection and typographic design remain two of the most important aspects of map design. As has always been the case, the type selected by the mapmaker should help the map emote feelings about the information presented and be an active conveyor of the message that is being communicated. Typography on Web maps is largely in its infancy. Many Web maps just use the default labeling techniques of GIS packages or map service application programming interfaces (APIs) and might therefore be why people still pine for well-crafted and -designed print maps. However, there is no reason for many of the labeling conventions and type standards of print cartography to be abandoned in Web mapping. The main thing to realize with typography and Web maps is this: You can design your Web maps with great typography, but unlike print maps, you will never have full control over how it looks on someone's computer or mobile device screen. Nevertheless, spending time thinking about your map's fonts and how they fit in with the overall message of your communication will definitely make your Web map more memorable.

Key Concepts

- Not all fonts are available to all users. Be sure to use fonts that are commonly available if you want slightly more control over how your map will look on most people's screens.
- You do not have to settle for default API labeling techniques. Many of the techniques and rules of labeling from print cartography stand firm.
- Never use more than two different fonts (i.e., typeface families) on a map. All other font differences can be created by manipulating the properties of the font used.
- Generally, you should avoid most serif fonts on Web maps. If you do use serif fonts, it is advisable they be specifically created for display on screens (e.g., Georgia).
- If no one is going to need to copy your label names, it is wise to make sure your label text is not selectable. It leads to less confusion and fewer accidental selections.

Further Reading and Resources

Web Sites on Web Typography

Sitepoint. For example, see the following article: Hume, A. (2005, December 9). The anatomy of Web fonts article. http://www.sitepoint.com/anatomy-web-fonts/ (short URL: http://goo.gl/kcQLk).

WebdesignerDepot.com. For example, see the following article: Gaines, K. (2013, February 12). 20 typographic Websites. http://www.webdesignerdepot.com /2013/02/20-typographic-websites/ (short URL: http://goo.gl/OG5gp).

Further Reading

Buckley, A. (2012, Summer). Designing great Web maps. *ArcUser*, 50–53.

Dooley, M., & Muehlenhaus, I. (2010). Mapping UFO sighting data: pitfalls and possibilities. *North American Cartographic Information Society Annual Conference.* October. St. Petersburg, Florida.

Jenny, B., Jenny, H., & Raeber, S. (2008). Map design for the Internet. In M. P. Peterson (Ed.), *International perspectives on maps and the Internet* (pp. 31–48). New York: Springer.

Slocum, T. A., McMaster, R. B., Kessler, F. C., & Howard, H. H. (2008). *Thematic cartography and geovisualization* (K. C. Clarke, Ed.) Prentice Hall Series in Geographic Information Science. Upper Saddle River, NJ: Pearson Prentice Hall.

Weinschenk, S. M. (2011). *100 things every designer needs to know about people* (p. 242). Berkeley, CA: New Riders.

7

Core Visual Variables

Clear and effective map design depends on your knowledgeable use of visual variables for the representation of your map data. Visual variables are graphic manipulations that symbolize data in a meaningful manner. Essentially, visual variables are used to convey information about the *nature* of data. Many of our abilities related to graphical interpretation are limited by evolutionary constraints. Our brains have been hardwired throughout time to interpret certain visual stimuli in particular ways. For example, humans perceive darker color values as representing more of something because in our environment, darker objects tend to be denser or weigh more (see Figure 7.1). Thus, light blue looks shallower than dark blue.

Just as pertinent to mapmaking, however, is that people also interpret graphics based on cultural norms and experience. For example, many of you probably assumed the blue colors were representing water in Figure 7.1. This is learned interpretation. This is also a clear example of a visual variable: The blue hue represents a particular map feature.

Defining the Visual Variables

The first cartographer to specifically address visual variable use in mapmaking was Jacques Bertin (Bertin & Berg, 1983). In *Semiology of Graphics,* he identified six core variables that are used in mapmaking and information graphics. When appropriately employed, he argued, it is these variables that allow us to make sense of information presented in graphic form (see Figure 7.2). When these variables are manipulated intuitively, people do not even need a legend to interpret the information displayed. The six core variables Bertin initially identified were shape, hue, value, orientation, texture, and size.

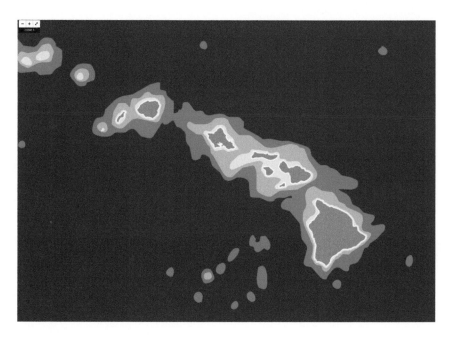

FIGURE 7.1
Dark blue is perceived as being deeper.

Over time, other cartographers began to scrutinize Bertin's six variables and identified additional ones. Although consensus on all of these additional variables is unattainable, by and large most of the variables found in Figure 7.2 (e.g., perspective height, value, saturation) are recognized and can be used when designing maps.

Shape

Shape is one of the most obvious methods of manipulating how an object will be interpreted. This is particularly true when shape is manipulated in such a way to mimic established cultural and cartographic norms. Figure 7.3 exemplifies this. Which of the point symbols do you assume is a capital city without even needing a legend?

Hue

Our lives are filled with color. Thus, our brain has grown adept at picking out different colors and associating them with different things; the color brown for many is associated with dryness, green with lushness. In Figure 7.4, which part of the map is land, and which is water? Even though this is not a real map, your brain will make an assumption based simply on the hues that are shown.

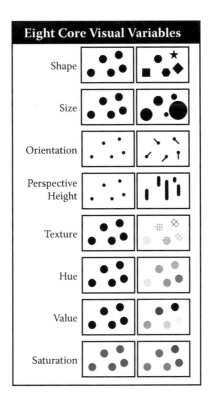

FIGURE 7.2
Visual variables.

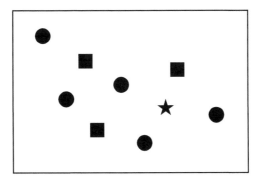

FIGURE 7.3
Shape.

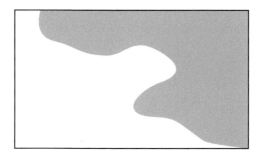

FIGURE 7.4
Hue.

Orientation

Graphical orientation can also help cue someone to differences between objects. For example, in Figure 7.5, can you make any conclusions about which cities are more right or left leaning in political orientation based on the results of a recent election?

Texture

Texture was very important for differentiating between objects before the advent of modern color printing. (Remember dot matrix printers?) In Web cartography, texture often comes in handy simply because it is so novel to see it used. In addition, texture remains one of the most effective ways to differentiate linear symbols (e.g., trails, bike paths, city streets, and highways).

Texture is often broken down into two more nuanced visual variables: spacing and arrangement. Spacing deals with how a texture or pattern is spaced (an example might be contour lines, as the closer together they are, the steeper an incline). Arrangement deals with how a texture is arranged compared to other similar features. Figure 7.6 shows two examples, one showing how different types of linear features might be created using arrangement and the other showing differences via line spacing.

Size

Differentiating between the sizes of things is intrinsic to humans. In maps, size is frequently manipulated to highlight differences in objects. In Figure 7.7, which is the largest city? Neither text nor data are needed. You make assumptions when size is manipulated.

Perspective Height

Perspective height is increasingly used on the Web to show differences in quantity. It consists of manipulating the heights of objects so that they can be

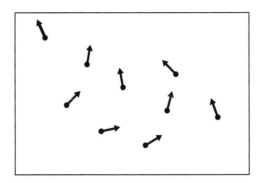

FIGURE 7.5
Orientation.

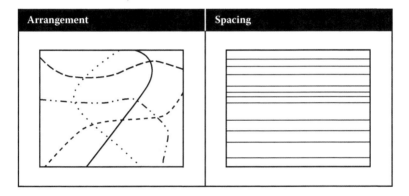

FIGURE 7.6
Texture.

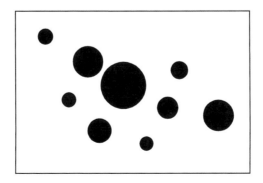

FIGURE 7.7
Size.

compared to one another. It is called perspective height because doing this often gives the appearance of an oblique or tilted perspective even though the data are being represented on a two-dimensional surface. Care should be taken to avoid having symbols near the front of your visualization cover symbols further behind. Also, this visual variable should never be used in a three-dimensional (3D) globe-like interface (e.g., Google Earth), as comparisons between data are almost impossible to interpret due to the curvature of Earth. Figure 7.8 shows an example of perspective height used to highlight differences in point values.

Value

As already discussed in Chapter 5, color value is the lightness or darkness of a hue. It is important to note that changing the lightness or darkness of a color's value will have an impact on how it is perceived. Darker values are interpreted as heavier and denser compared to lighter-color values. In the map in Figure 7.9, which Australian states do you presume have the highest rate of green bean consumption?

Saturation

Also discussed in Chapter 5, color saturation represents how bright or intense a hue is. A color can range in saturation from a dull gray (gray scale) to the full hue itself (full saturation). All colors can be described using hue, value, and saturation; yet, all three of these variables can be manipulated individually to create map symbols. Typically, saturation is manipulated in conjunction with value on maps. Figure 7.10 shows the impact of changing a map symbol's color saturation.

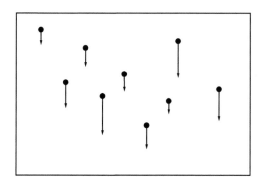

FIGURE 7.8
Perspective height.

FIGURE 7.9
Value.

However, not all visual variables serve all data types equally well. Certain variables are only appropriate for showing particular data types. In fact, if one uses inappropriate visual variables, the meaning and accurate interpretation of the map data may be lost.

Visual Variables for Mapping Qualitative Data

Qualitative data represent categories. Qualitative spatial data tell us what and where things are. Some common examples of qualitative data found on a map include countries, land cover, land use, language diffusion, and ethnic groups. These things have no quantitative or comparable values in and of themselves. If your data are qualitative in nature, the following visual

FIGURE 7.10
Saturation.

variables are appropriate: shape, hue, orientation, and texture (arrangement). Figure 7.11 provides examples of how these variables can be used for different types of point data.

Visual Variables for Mapping Quantitative Data

Quantitative spatial data are associated with a value of some sort. Quantitative data tell how much (or at what level) something occurs. Several types of quantitative data exist: interval, ratio, and ordinal data.

Data referred to as interval data convey quantitative measurements using an arbitrary zero. Data can be compared to other data in a numeric fashion, but the zero point is largely meaningless. Examples of interval data would be temperature readings in degrees celsius or elevations in meters above or below sea level.

Ratio data are quantitative data that have absolute zeros; zero means nothing is present. Percentages are always ratio data, as are measurements such as population and rainfall. You either have it or not.

Finally, some quantitative data are not metric in nature; that is, specific measurements of difference between data are not measurable. All that one has is an order of magnitude for the data. These are referred to as ordinal

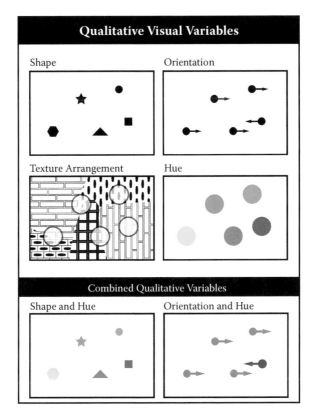

FIGURE 7.11
Visual variables that can be used to show qualitative data.

data. An example of an ordinal dataset would be a ranking of the happiest countries in the world. You cannot tell how much happier Danes are than Argentinians, but you can tell their difference in ranking (a higher ranking vs. a lower ranking).

If your data are quantitative in nature, the following visual variables can be used to represent your data: size, value, perspective height, orientation, and texture (spacing). Figure 7.12 provides examples of how these variables can be used for different datasets.

Combining Visual Variables

Combining several of these visual variables will typically result in more intuitive map symbols. It is often a good idea to reinforce the meanings of symbols by using more than just size, shape, or color schemes to differentiate your map objects. Often, we do this without even thinking. Figure 7.13 shows how this could be done with several of the previously presented visual variable examples.

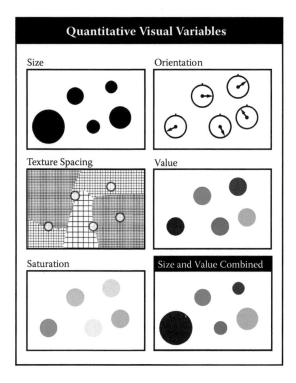

FIGURE 7.12
Visual variables that can be used to show quantitative data.

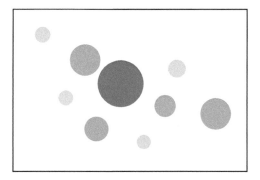

FIGURE 7.13
Combination of visual variables (size and color value).

Again, map communication is done through the use of these visual variables. Your base map, your point symbols, and even your thematic symbols are designed by manipulating these different graphical attributes. Effectively deciding how and when to use these variables, particularly with one another, will help to convey your map's message clearly.

- The design of maps is largely shaped via the manipulation of visual variables. These visual variables include shape, size, perspective height, orientation, texture (both via pattern spacing and arrangement), hue, value, and saturation.
- Certain visual variables are best used to highlight qualitative data (i.e., shape, hue, orientation, and texture arrangement).
- Other visual variables are better for quantitative data (i.e., size, value, perspective height, orientation, and texture spacing).
- Knowing when to use particular visual variables is important for designing effective maps.
- Often, it is wise to combine several visual variables into a single symbol to emphasize and reinforce the meaning of the symbol.

Further Reading

Bertin, J., & Berg, W. J. (1983). *Semiology of graphics* (p. 415). Madison: University of Wisconsin Press.

Dondis, D. A. (1973). *A primer of visual literacy* (p. 194). Cambridge, MA: Massachusetts Institute of Technology.

Hoffman, D. D. (1998). *Visual intelligence* (p. 294). New York: Norton.

MacEachren, A. M. (1994). *Some truth with maps: a primer on symbolization and design. Resource publications in geography* (p. iv, 129 pp.). Washington, DC: Association of American Geographers.

MacEachren, A. M. (1995). *How maps work: representation, visualization, and design* (p. xiii, 513 pp.). New York: Guilford Press.

Monmonier, M. S. (1993). *Mapping it out: expository cartography for the humanities and social sciences. Chicago guides to writing, editing, and publishing* (p. xiii, 301 pp.). Chicago: University of Chicago Press.

Slocum, T. A., McMaster, R. B., Kessler, F. C., & Howard, H. H. (2005). Thematic cartography and geographic visualization. In K. C. Clarke (Ed.), *Prentice Hall series in geographic information science* (2nd ed., p. 518). Upper Saddle River, NJ: Pearson Prentice Hall.

Tufte, E. R. (1991). *Envisioning information* (p. 126). Cheshire, CT: Graphics Press.

8

Symbolization

Introduction

Unfortunately, what might pop into many people's minds when they hear "Web map" is a map full of upside-down teardrops, pushpins, or thumbtacks (see Figure 8.1). These default symbols largely rose to prominence on the Web with the advent of Google Maps and Google Earth.

The keyword that needs to be highlighted in the previous paragraph is *default*—as in basic, nonspecialized, even simplistic. There is nothing inherently wrong with someone who is untrained in map design creating a Web map of all the places he or she has been in Arkansas using default thumbtack symbols. Not everyone has the time or inclination to learn Web mapping software, much less read about effective Web map design. However, assuming you are reading this book because you want to design good intuitive Web maps, it is probably wise to start thinking of default map symbols as nothing more than temporary placeholders.

As Web map designers, we must remind ourselves of the role visual variables play in graphical literacy. It is extremely rare that default Web map symbols adequately portray the information they are representing in an intuitive manner. They thereby interfere with the goals of Web map communication. The default symbol, such as an upside-down teardrop, tells the map user nothing about a mapped feature itself. The map user will need to refer to a title, legend, or the text of a Web site to determine what such a symbol represents. Effective visual communication demands intuitive symbolization.

Designing Effective Web Map Symbols

Map symbol design is an enormous topic. However, without spending hundreds of pages covering the intricate theories of symbol design, there are a handful of strategies that I highly recommend when designing symbols for any given Web map project.

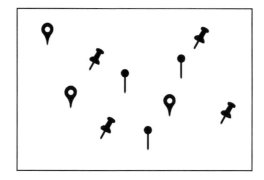

FIGURE 8.1
Thumbtacks, inverted teardrops, and pushpins: the poster children of Web maps.

Keep Symbols Simple

One thing to remember when designing map symbols for the Web is that many screens have poor resolution. Not only that, but when designing point symbols for your Web map, much of the time, your symbols will be quite small. Thus, minimize the amount of intricate detail on your map symbols. Not only is it likely that most people will not notice the fine detail you have spent time on, but also because symbol details often become so distorted on the Web, the symbol itself looks awkward and is distracting (Figure 8.2).

Generic Caricatures Are Best

Whenever possible, avoid creating symbols that may accidentally imply details about a place that are not necessarily accurate. For example, if you are designing a symbol to represent a dog park, you should try to make the symbol inclusive to all dog breeds and mutts. Although when you hear "dog" the first breed that pops in your head may be a German shepherd, this will not be the case for everyone. So, if you use a symbol that looks a lot like a German shepherd for your dog parks, someone who has a Wheaten terrier that was attacked by a German shepherd at a dog park may wince. Rather than create a symbol looking specifically like a German shepherd, it is best to try to create as generic a dog image as possible (Figure 8.3).

Symbols Are Era and Audience Dependent

Symbology is constantly evolving. When I was growing up, a television store might have been represented on a map as shown in Figure 8.4a. Today, a television store might better be represented as shown in Figure 8.4b. How about telephones? Gone are the rotary phones represented in Figure 8.4c and in fashion are those in Figure 8.4d. Or, are they? Perhaps with the advent of smartphones, this also will change (Figure 8.4e).

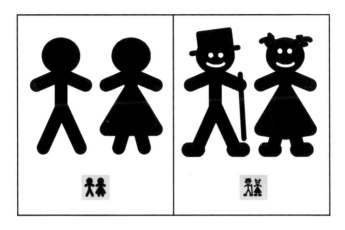

FIGURE 8.2

On the left: Men and women symbols, perhaps representing restrooms. On the right: Men and women symbols, perhaps representing a restroom at the opera? Notice that the more intricate symbols do not resize nearly as well.

FIGURE 8.3

On the left: generic dog symbol for a dog park. On the right: German shepherd symbol for a dog park. Both resize adequately.

Will most children being born today ever understand the symbols found in Figures 8.4a and 8.4c? It depends. If these symbols are continually used and become ubiquitous, they may become icons imbued with meaning, even meanings they were not originally meant to represent. For example, the symbol for circuit breaker is now universally recognized as the on/off symbol (see Figure 8.4f). Regardless, when you are designing symbols, make sure that you are not using antiquarian icons. At the same time, if certain mapping conventions are known to exist, do not design new symbols just for the sake of it. Use people's knowledge of symbols to your advantage.

Symbols Are Context Dependent

Symbols will have different meanings depending on the nature of symbols around them. A classic example I like to use in class is a CIA map published in 2002 that shows Iraq's purported possession of weapons of mass

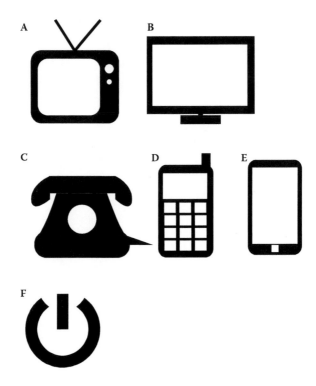

FIGURE 8.4
Different symbols as referred to in the text. Symbol A: analog television from the 1980s. Symbol B: modern television. Symbol C: rotary telephone. Symbol D: mobile phone. Symbol E: (mobile) smartphone. Symbol F: power button symbol.

destruction. Regardless of how inaccurate the data represented on the map were, the most egregious error on this map was the choice of symbols. Please look at the map in Figure 8.5 before reading further. This is a re-creation of the main portion of the original CIA map. Which of the symbols represent weapons that the United States then knew no longer existed? Symbol context matters as much for accurate interpretation as the symbol itself.

How Do You Represent a Stadium?

In 2011, at the North American Cartographic Information Society Annual Conference (http:///www.nacis.org) in Madison, Wisconsin, I saw one of the most enlightening presentations of my life. The speaker was Patrick Hofmann, who was at the time working for Google. He had done a lot of work on the design of Google Maps' symbols. (Not the pushpins, but the actual point symbols used on Google Maps to represent different things, such as restaurants, places of worship, and stadiums.) During his talk, he discussed how difficult it is to design effective symbols that are universally

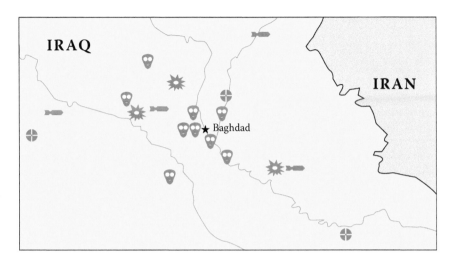

FIGURE 8.5

Map excerpt from *Iraq's Weapons of Mass Destruction Programs* report (CIA, 2002). This was re-created here as the original version is low resolution. The explosions represent weapons that the United States knew did not exist. If a map user did not look at the legend, the user might be forgiven for missing this important information. (Available at https://www.cia.gov/library/reports/general-reports-1/iraq_wmd/Iraq_Oct_2002.pdf)

recognized. One example he provided was the challenge of designing a symbol to represent stadiums. Google's stadium symbol needed to be recognized globally. It had to be applicable to all stadiums and all arenas (covered sports facilities), regardless of the type of sports played there. Take a second to go to Google Maps and search for several stadiums of different sports, such as the Aker Stadion in Molde, Norway, and MetLife Stadium in East Rutherford, New Jersey. Check out the symbol Google decided to use. This example reinforces the concept that simple and less-descriptive symbols are often better for universal understanding of map data.

Key Concepts

- Map symbols are almost always more effective when they are simple and pictographic in nature.
- Symbol interpretation and perception depend on a variety of audience and map variables. Symbol meaning and identification vary depending on era, audience, and map context.
- Never, ever, use ambiguous, default Web map symbols (e.g., upside-down teardrops, pushpins, or thumbtacks) unless your client insists on it. Even then, try to get out of it.

Further Reading

CIA. (2002). Iraq's weapons of mass destruction programs. Retrieved from https://www.cia.gov/library/reports/general-reports-1/iraq_wmd/Iraq_Oct_2002.pdf

Dondis, D. A. (1973). *A primer of visual literacy* (p. 194). Cambridge, MA: Massachusetts Institute of Technology.

Hoffman, D. D. (1998). *Visual intelligence* (p. 294). New York: Norton.

MacEachren, A. M. (1994). *Some truth with maps: a primer on symbolization and design. Resource publications in geography* (p. iv, 129 pp.). Washington, DC: Association of American Geographers.

MacEachren, A. M. (1995). *How maps work: representation, visualization, and design* (p. xiii, 513 pp.). New York: Guilford Press.

Monmonier, M. S. (1993). *Mapping it out: expository cartography for the humanities and social sciences. Chicago guides to writing, editing, and publishing* (p. xiii, 301 pp.). Chicago: University of Chicago Press.

Roth, R. E., Finch, B. G., Blanford, J. I., Klippel, A., Robinson, A. C., MacEachren, A. M., Vi, C., et al. (2008). The card sorting method for map symbol design. 2010 International Symposium on Automated Cartography (AutoCarto) July, Paris, France. Retrieved from http://www.geovista.psu.edu/publications/2010/RothEtAl_2010_AutoCarto.pdf

Wallace, T. (2011). A new map sign typology for the geoweb. *Proceedings of the 25th Annual International Cartographic Conference.* Retrieved from http://icaci.org/files/documents/ICC_proceedings/ICC2011/Oral%20Presentations%20PDF/E1-Symbols,%20colors%20and%20map%20design/CO-417.pdf

9

Thematic Visualization

Introduction

Thematic cartography at its most basic is simply the graphic symbolization of spatial datasets concerning a particular theme over a base map. There are numerous types of thematic representations, or map styles, that can be used for spatial visualization (e.g., choropleth, dot, proportional symbol). Regardless of which form of representation is employed, all thematic maps are created by manipulating visual variables.

As noted in Chapter 7, most visual variables are only ideal for representing certain types of data. Thus, one of the first things to remember is that *not all thematic representations are suitable for mapping all types of data.* Be wary of using visual variables that do not accurately, or intuitively, highlight the nature of the information they are presenting.

This chapter is devoted to helping people who are new to cartography learn more about the different types of thematic representations available to them. This chapter is also meant to help those who have a background in cartography better understand the idiosyncratic benefits and drawbacks of using different thematic representations in their Web maps. (There are several new hurdles confronting traditional representation techniques.) The next section reviews data and technology issues to consider when designing thematic maps for the Web. This is followed by a review of the most popular types of thematic representation. I also discuss the benefits for each type of representation. The chapter concludes with a brief discussion about the advantages and disadvantages of allowing map users to interact with and manipulate the thematic data that are represented.

Data and Technology Limitations

When designing thematic maps for the Web, mapmakers are immediately confronted with two hurdles: (1) the nature of the data being mapped and (2) the limitations of the technology used to create the map.

The Nature of the Mapped Data

The spatial data you are attempting to map will vary depending on many things, but nothing is more influential than the nature of the phenomenon being sampled and how the data were collected.

Cartographers map data. Data are not real. Data are measurements or qualitative samples taken from a social or environmental phenomenon. For example, if we were to map population density in the metro area around Adelaide, South Australia, we would not actually be mapping the current population. We would be mapping data collected by a census and therefore a snapshot of reality. Data are almost always based on samples from a phenomenon—in this case, the metro population of Adelaide. Even when they are not sample based, data are time specific. When we get around to actually mapping the data, we must assume the phenomenon has likely changed to some extent.

Different phenomena have different spatial properties (MacEachren, 1995). Some phenomena occur discretely, at what we might describe as particular places. Others occur continuously across space (see Figure 9.1). For example, zoos are a discrete phenomenon. They are not found continuously across the

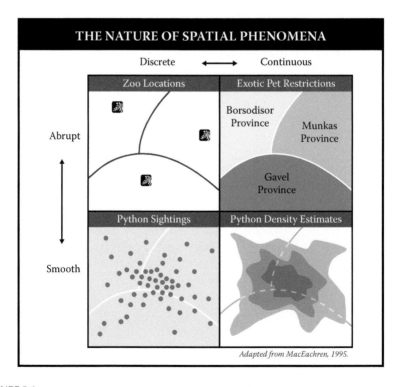

FIGURE 9.1

Diagram of phenomena characteristics. (Adapted from MacEachren, A. M. 1995. *How maps work: representation, visualization, and design*, p. xiii, 513 p. New York: Guilford Press.)

landscape. On the other hand, exotic animal laws and regulations (e.g., the legality of Burmese python sales) are a continuous phenomenon. They exist across the landscape and change only at state or provincial borders.

Phenomena can also be categorized as abrupt or smooth (see Figure 9.1). Zoos are abrupt. They not only occur in discrete places but also are randomly placed across a landscape, and one can go hundreds of kilometers without finding one. Zoos rarely cluster; one only needs so many tigers in a city. Exotic animal laws and regulations are also abrupt. As soon as I drive across the border from Florida into Alabama, the laws and regulations could change. The number of Burmese pythons currently slithering around in the wilds of Florida right now, on the other hand, is an example of a smooth phenomenon. Their numbers smoothly dissipate as one retreats from the center of their ideal habitat.

The method of data collection about phenomena will have an impact on how we can map them. A phenomenon can be smooth, like python distribution in Florida, but we might only be able to obtain abrupt-continuous data about them (e.g., county-level data regarding how many wild pythons are estimated to be therein). While it is always good practice for cartographers to use data that are as close in nature as possible to the phenomenon being mapped, this is often impossible. When it is possible, then it is important to realize the limitations of your representation and what impact these limitations may have on your communication goals.

Technological Limitations

The second issue all Web cartographers must deal with in thematic cartography is technological constraints. Which software, techniques, or application programming interfaces (APIs) you decide to use to design your thematic map will affect the types of representations you can, or should, make.

Given certain Web mapping technologies (discussed in more detail in Chapter 12), it is sometimes difficult or even impossible to use the most appropriate thematic representation (e.g., an isarithmic map instead of a choropleth one) for your map data. Thus, people often settle for a representation technique that is available to them rather than look for technology that allows them to represent the data using a more insightful method. Settling for an inappropriate representation will likely misguide your map users and cause them to come to incorrect conclusions concerning your mapped theme.

Mercator's Apparition

There is a specter haunting Web mapping. It is the Web Mercator projection. One technical problem every developer eventually confronts is this projection's ubiquity. The Web Mercator projection, misused as it is, is the default projection of the Web mapping world. While it is actually a fine projection for large-scale maps or maps designed for several types of navigation, it is an absolutely horrible projection for medium-to-small-scale thematic maps.

The Web Mercator grossly distorts the area of places not found around the equator. Why does distortion matter? A prime example is found in Figure 9.2. This map distorts how much of the world is heating up and where. The sizes of Arctic and Antarctic places are far too large. (In reality, Greenland has roughly the same areal extent as Mexico.) This may be fine if the purpose of your map is to convince an audience that they must take action to prevent the oceans from rising. Indeed, it provides a rhetorical advantage. However, if you are trying to give an accurate representation of the impact of climate change for students or scholars, a Mercator projection can be seen as unethical. Some APIs allow you to change the projection. If Web Mercator is your only option with a Web mapping service or API, it is often best to switch technologies. Whenever possible, you should always use an equal-area projection for your thematic maps.

Different Thematic Representations

Over the past 200-odd years, numerous types and styles of thematic representation have been developed. There is no way that a single chapter can do justice to all of the options available to any cartographer. The goal of the

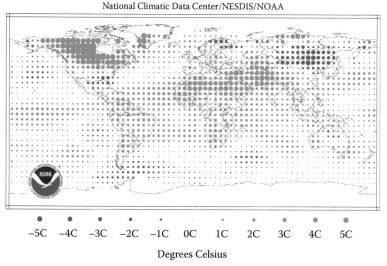

Temperature Anomalies March 2010

(with respect to a 1971–2000 base period)

National Climatic Data Center/NESDIS/NOAA

−5C   −4C   −3C   −2C   −1C   0C   1C   2C   3C   4C   5C

Degrees Celsius

FIGURE 9.2

Map created by NOAA that grossly distorts the size of the polar regions.

rest of this chapter is to give you a cursory overview of the most common representations used in cartography. Each thematic style is described; the types of data (e.g., abrupt-continuous) that can be effectively communicated are discussed, and the advantages and disadvantages of using each method are highlighted.

Choropleth Maps

As the Mercator is to Web projections, choropleth maps are to thematic cartography (i.e., they are overused). Choropleth maps use different values, hues, or saturation to show data differences among enumeration units (e.g., provinces, counties, countries). When used appropriately, the choropleth technique is extremely effective for intuitively displaying spatial information. Unfortunately, choropleth maps have several shortcomings that are not always realized by mapmakers.

Before we go any further, it is imperative we outline a few issues confronting the effective use of choropleth maps in particular. First, if you have raw data values (i.e., count data), *do not* use a choropleth map. Choropleth maps should only be used to represent standardized data (e.g., percentages, rates, people per square mile). By mapping raw data with a choropleth map, you fail to take into consideration the impact that different size enumeration units will have on a value. For example, let us analyze differences in population within the German federal states. If we map raw population data for each state in choropleth form, we obtain map at left in Figure 9.3. However, this is a bit misleading. The states vary drastically in area. States with larger areas would be expected to have higher population totals. Map on right maps population standardized by state area (i.e., the population densities of each state). The maps communicate almost opposite messages.

Second, generally all thematic maps should be created using equal-area projections, but there are additional reasons this is even more important when it comes to choropleth maps. Due to the fact that choropleth maps fill entire enumeration units with a color value, areal distortion can massively distort not only the nature of the data being displayed (e.g., how big places are) but also the interpretation of the overall dataset. A proportional symbol (discussed further later in this chapter), for example, will sit in the middle of an enumeration unit no matter how large it is. But, by not using an equal-area projection for a choropleth map, many valid patterns may be deemphasized or irrelevant ones overly embellished.

Finally, value and saturation are the ideal visual variables to use when designing a choropleth map. Referring back to Chapter 7 on visual variables, value and saturation imply quantitative values. Hue, on the other hand, does not. If you design a choropleth map using only hue, your map may be difficult to interpret. There are only a limited number of hue transitions that work to represent differences in quantity (e.g., yellow to red). Most hue changes need to be coupled with changes in value and saturation. Other than those hue

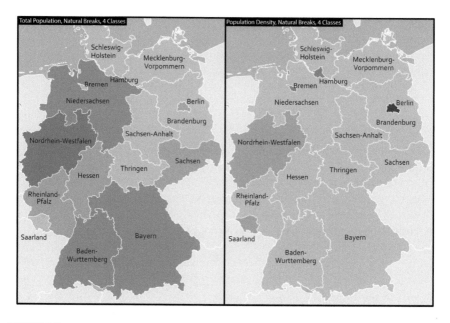

FIGURE 9.3

Left: A choropleth map of population count data. Right: A choropleth map of population density. Notice how misleading the results are when one does not use standardized data.

combinations found on Color Brewer (discussed and referenced in Chapter 5), never use hue to represent data on maps representing quantitative values.

Data Classification

Most choropleth maps classify data. Data classification is the process of analyzing the distribution of values in a dataset and grouping them based on their location in a histogram. Each group is then assigned a color specification for representation on the map. Choropleth maps should typically have between three and five classes. In most cases, using more than five classes makes it nearly impossible for humans to differentiate between the colors found on a map. There are numerous ways of classifying one's data. Next, I review some of the most common methods.

Natural Breaks

One of the most commonly used classification techniques is the natural breaks method. Using this method, the largest gaps between data values in a histogram are marked, and the resulting groups of values are shaded accordingly (Figure 9.4). This way, the largest gaps in the dataset come to represent the breaks between classes. This is often a useful method for highlighting natural clustering within your dataset. It can be particularly useful if you have several high or low outlier values, as they will typically be classed separately. Figure 9.4 shows an example of using natural breaks to classify the population densities of the federal states of Germany.

Quantiles

The quantile method of classification takes the total number of data values found on a histogram and divides that by the desired number of classes. The result is that each class contains the same amount of data values.

FIGURE 9.4

Natural breaks choropleth map of German population density.

For example, if you were mapping the population density of the 16 federal states of Germany, you would place these values on a histogram (Figure 9.5). You would likely decide on four classes to divide these states into four different color values on your map. (Quantiles with four classes are referred to as *quartiles*.) Sixteen divided by four would result in four federal states being included in each class. You would then start counting up from the bottom of the histogram, creating a new class for every four provinces (Figure 9.5). The quantile classification method should generally only be used for ordinal-level data, that is, data that are ranked, not measured.

FIGURE 9.5
Quantiles choropleth map of German population density.

Using it for German province density would be potentially misleading, as several provinces that are very close to one another in density values would be separated by an arbitrary class break.

Equal Intervals

Equal-interval classification is another commonly used method of data classification. One simply takes the total range of data values and divides by the number of classes desired (see Figure 9.6). One is left with classes that

FIGURE 9.6
Equal-intervals choropleth map of German population density.

possess an equal data range. Some of your classes may have many more data values than others. One should avoid creating classes that contain no data by either increasing or decreasing the number of classes used.

Standard Deviations

The aforementioned classification schemes are stellar if you are trying to show data values in a hierarchical fashion—from lowest to highest or least to most, for example. However, sometimes we want to highlight how data vary from a central point or average. The standard deviation classification scheme allows us to do just this. This scheme requires that you compute the mean and standard deviation of your dataset. Once this is known, you can use a diverging color scheme to highlight whether a federal state in Germany is above or below the average of all provinces. Unfortunately, this classification scheme is really only optimal for normal distributions, which are infrequently found with geographic data.

Unclassed Choropleth Maps

Finally, there are also unclassed choropleth maps (Figure 9.7). Unclassed choropleth maps give each enumeration unit a different color value based on where it lies on the histogram. Unclassed choropleth maps are heralded by some as better than their classed peers because they do not generalize the data. Instead, it is argued, they allow a map user to view differences between specific enumeration units. Many, however, disagree with this notion. The counterargument against unclassed maps is that it becomes very difficult, and downright impossible, for humans to discern the difference between many more than five different color values. Thus, a map that has 16 or 60 different color values will overwhelm the cognitive capabilities of map users.

Although I tend to concur with the latter group, these maps can be enlightening if you are trying to reveal areal patterns in your dataset. Often when you are designing a choropleth map, you may want first to create an unclassed map to see if any areas have higher or lower values than you might expect. This can help you choose an appropriate classification scheme for presenting your information.

Interactivity has also made unclassed maps more useful. One no longer has to attempt to distinguish the color value difference between two enumeration units; one can simply select the two enumeration units and, if an info window with the data is provided, determine the difference immediately by looking at the actual values. To some extent, the argument against using unclassed maps is diminishing with interactive Web cartography.

When using an unclassified choropleth map to show differences from a central point (e.g., a median or mean), use a diverging color scheme (Figure 9.8). Diverging color schemes consist of a neutral color in the middle,

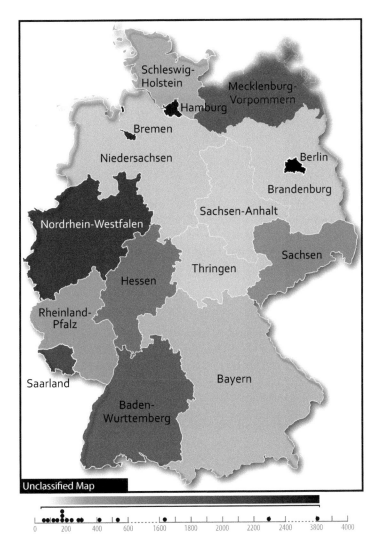

FIGURE 9.7
Unclassed choropleth map of German population density.

with two different hues at either end. Color value and saturation are manipulated between the middle neutral color and the two end colors. These can be intuitive to read. Avoid using a rainbow color scheme with unclassified maps as this can be easily misinterpreted.

Benefits of Choropleth Maps

There are some serious benefits to using a choropleth map design. First, they are easy to make. Second, when designed appropriately—with lighter color

Diverging Color Scheme

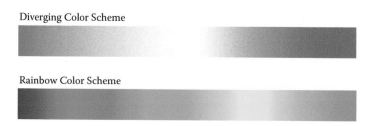

Rainbow Color Scheme

FIGURE 9.8

An example of a diverging color scheme and a rainbow color scheme. Avoid rainbow color schemes in general as they are not as intuitive as well-designed diverging color schemes.

values representing less of something and darker color values representing more—they are extremely intuitive. Finally, choropleth representations are excellent for mapping data that represent continuous-abrupt phenomena (please refer to Figure 9.1).

Drawbacks of Choropleth Maps

Choropleth maps are often used to map data that might be more appropriately mapped using a different technique. Choropleth maps are notorious for covering up important information about the distribution of data *within* an enumeration unit. For example, population density is not equal throughout the different German federal states. Yet, by filling in polygons with a solid color, the choropleth maps represent and reify each federal state as equally dense. A dot map would be far more efficient for showing more detailed population distribution within Germany.

Dot Maps

Dot maps are excellent representations for showing the spatial distribution of datasets within enumeration units. Dot maps use point symbols (most frequently "dots") to represent values. A dot may either reference individual objects or may be conglomerated to represent a predetermined number of objects (see Figure 9.9). Dot maps are most effective at showing discrete data, both abrupt and smooth.

Though dots maps traditionally use their namesake symbol, two dot styles exist: geometric and mimetic. Geometric symbols (e.g., squares, dots, triangles, etc.) are effective when you have many values conglomerated close together. They tend to take up less space and to be easier to differentiate when densely packed. Mimetic symbols (also known as pictographic symbols) are those that mimic an image of what they are representing. For example, one might fashion a symbol like a soccer ball to map the distribution of public and recreational pitches and playing fields.

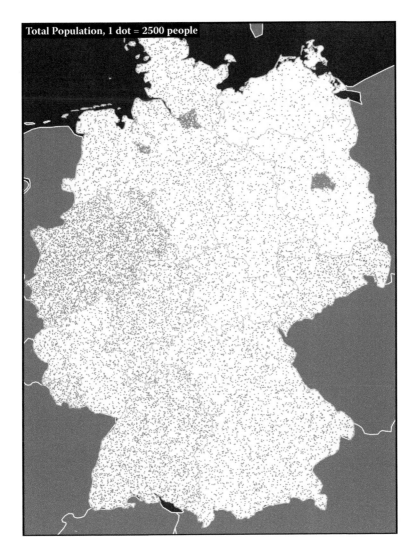

Total Population, 1 dot = 2500 people

FIGURE 9.9

Dot map of Germany's population (2010). Each dot represents 2,500 people. Dot maps provide map users the ability to get a sense of both a density and a total count of the data being mapped. A 1:1 dot-to-person ratio for population is impractical in this situation.

Heat Maps

Heat maps are a hybrid form of dot maps that are frequently used on the Web to display the spatial intensity of point data (see Figure 9.10). Heat maps get their name from the fact that they show data using a hue range from blue (a cool color representing fewer) to green (a neutral color representing a modest amount) to red (a warm color representing more). When many data

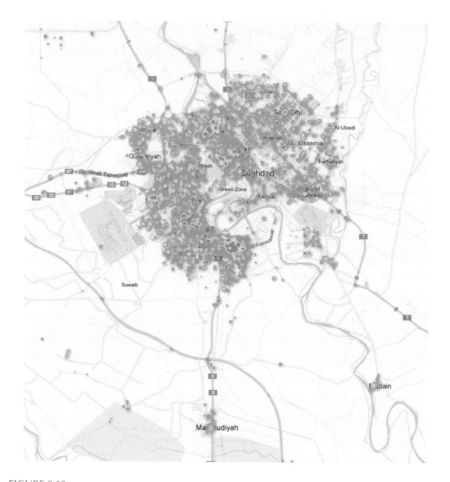

FIGURE 9.10

Heat map showing U.S. casualties in the area around Baghdad, Iraq, through 2008. (Created using Google Fusion.© 2013 Google. With permission.)

points are densely packed, the area is represented with a more reddish color. In areas where data points are still abundant but not as densely concentrated, green is used. In areas with sparse data, blue is used. Finally, areas with no data are not represented at all; they are simply left blank.

Heat maps do not show dots per se. Instead, they show where point data occur by placing buffers around the points and coloring the resulting polygon. As point areas overlap, the density increases, and the color shifts from blue toward green and then red.

Although visually quite exciting to look at, and easy to make using a variety of online mapping tools, heat maps have many drawbacks. First, it is impossible to know visually how many points there are in a particular area. Heat map legends can help with this, but heat maps necessarily make a mess of point data and do not let one discern exact locations of the points being plotted. Second, heat maps are sometimes used to map data that are not intuitively associated with a blue-to-red color spectrum. For example, if you are mapping incidents of violent crime, a map user may perceive both blue and red areas as prone to violence. (As mentioned in the chapter on color, blue is often associated with negativity and depression, whereas red is associated with aggression and danger.) By default, even with a legend, this makes the map more difficult to interpret.

Benefits of Dot Maps

Dot maps are often easy to create. If you have specific data points (e.g., latitude and longitude, geocoded addresses), it is often a fairly easy endeavor to import and map them. If you have data for artificial collection or enumeration units (e.g., townships and municipalities, counties, zip codes), it is often easy to determine a reasonable dot value and dot size to intuitively distribute dots within these units.

Drawbacks of Dot Maps

Dot maps are awesome—but only when the data are suitable for representation using this method and the symbols chosen are neither bloated pushpins nor upside-down teardrops. Just because your data are point specific does not mean they should be mapped using a dot map.

Dooley and Muehlenhaus (2010) explored the downside of using dot maps for point data by designing a series of UFO sightings maps. They hypothesized that contemporary geocoded dot maps of UFO sightings found on the Internet are not only unintelligible due to the overlapping dots, but also fail to display the true nature of the data. Instead, all such dot maps do is highlight the fact that there are more people living in cities, as more people see UFOs in cities than anywhere else. They demonstrated that when mapping UFO sightings data using proportional symbols or choropleth representations, new information emerges. It turns out that people in the Rocky Mountain region of the United States stumble upon far more UFOs than people living elsewhere in the United States. This information was completely nonexistent on dot map versions of the same data. Dot maps are powerful, but they are not infallible.

Proportional and Graduated Symbol Maps

Proportional and graduated symbol maps are excellent for visualizing the raw values of datasets tallied by enumeration unit (e.g., number of registered chiropractors by country). Like dot maps, proportional and graduated symbol maps use a symbol to represent values (most commonly a circle). Unlike dot maps, only one symbol per enumeration unit is needed (e.g., one symbol per country). Each symbol's area is resized in comparison to other values in the dataset.

The main thing to remember when designing proportional symbols is that you must size the symbols proportionately based on area, not radius, circumference, width, height, or any other form of measurement. If you do not make your symbols proportionate by area, you will end up with massively distorted sizes that grossly embellish differences in values.

The difference between proportional and graduated symbols is quite simple. Proportional symbols are sized proportionately to the other values in the dataset (see Figure 9.11). Thus, the symbol representing Bavaria

FIGURE 9.11
Proportional symbol map of Germany's population.

with its 12.5 million residents will be five times as large as the symbol for Brandenburg with its 2.5 million residents. Theoretically, this allows people to compare differences in values visually between enumeration units by looking at the size of the symbols. Realistically, humans have a hard time accurately perceiving differences in graphical areas, so map users will never truly be able to do this. However, by viewing symbols of different sizes, they will be able to obtain a nuanced idea of value changes between places.

Graduated symbol maps classify their data as choropleth maps do. A histogram is created of the data, and the data are classed into different groups based on their values using one of several classification methods (as discussed in the choropleth map section). Ideally, between three and five classes are created. Discernible symbol sizes are chosen for each class, resulting in a map that shows only a handful of different size symbols representing all values. The downside of graduated symbols is that a person will not be able to determine exact values by looking at the symbol. On the other hand, the map user will not be inundated with many different sized symbols. The user can quickly peruse the map and know if a place falls within a certain data range or not.

Benefits of Proportional and Graduated Symbol Maps

Quite simply, proportional symbol maps are fabulous. Size difference is one of the most intuitive visual variables for humans to comprehend. Therefore, these maps are extremely intuitive and easy to read. Proportional and graduated symbol maps are most effective at showing abrupt data that are discrete or continuous. They can be used to map raw data (e.g., number of chiropractors) or standardized data (e.g., the number of chiropractors per square kilometer), although general consensus seems to be that raw data are best. When the enumeration units are small enough, like dot maps, proportional and graduated symbol maps provide a view of a dataset's distribution across the landscape. Moreover, symbols can easily be placed over choropleth representations to add a second layer of data to a map. You can use not only geometric symbols but also mimetic symbols to add aesthetic zeal to the map's message. They are also remarkably easy to create with little more than a spreadsheet program.

Drawback of Proportional and Graduated Symbol Maps

Depending on the nature of your dataset, proportional and graduated symbols do have a serious drawback. If your enumeration units are relatively compact and small, proportional symbols tend to overlap. Symbol overlapping can result in data that are difficult to see and interpret. Due to their compact shapes, the use of circle symbols can help minimize potential overlapping. However, humans have an extremely difficult time interpreting the areas of circles; we do much better with squares. This is something to consider when deciding which shape to use.

Isarithmic Maps

Isarithmic maps represent data using lines. Each line on an isarithmic map represents a constant value occurring perpendicular to Earth's surface (Figure 9.12). The values of these lines are consistent (e.g., each distinct line on a topographic map represents 10 m in elevation). Most of us are familiar with this concept from weather maps, which often show areas of different temperatures using isotherms. (Note: There are dozens of specific names given to isarithmic lines depending on the type of data they are mapping; isotherms are lines representing temperature; isobars are lines representing atmospheric pressure; isobaths represent depth, and so on.) Hue, value, and saturation can

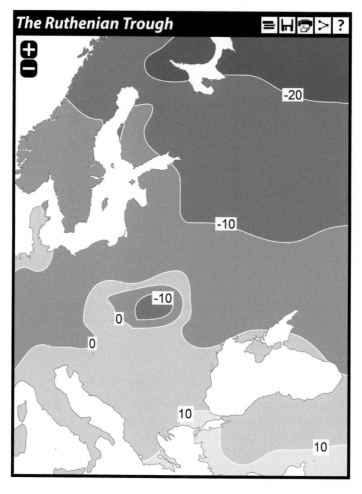

FIGURE 9.12
Isarithmic map.

also be used to highlight differences in isarithmic values on the map (e.g., iso-therm lines are frequently highlighted with a blue-to-red color scale).

There are two types of isarithmic maps that can be created depending on the nature of your data sample: isometric and isoplethic. Isometric maps are created from data collected at specific locations with x-y coordinates of some sort (i.e., point data). Examples of these types of data might be depths (collected via sounding weight), temperatures (collected at weather stations), and annual attendance at professional sports games (collected at stadiums). All of these data would be collected at points with a specific, nonpolygonal location. Isoplethic maps, however, represent values collected by enumeration unit, or polygonal area. These two types of isarithmic maps need to be constructed slightly differently.

Isometric maps are constructed using linear interpolation. Linear interpolation is the process of determining the placement of an isoline among data points. One determines the value difference between two control points and then draws the isoline through these two control points. The isoline is thus a best guess regarding the location of the actual value it represents. Of course, this means that isometric lines are not infallible as they are estimates.

Isoplethic maps need to be constructed differently because one cannot use linear interpolation for polygons. Thus, the first decision a cartographer must make when creating an isoplethic map is where to create default control points for the enumeration units. A variety of methods for doing this exists. Often, a point with a data value is simply placed at the centroid of each enumeration unit. Another technique is to pick a prominent point feature that lies within each of the polygons (e.g., capitals, largest cities, point of highest elevation). Regardless of the method you choose, it is important that you be consistent in determining your enumeration unit points. For example, do not choose a capital city for one province and the largest city for another. Finally, isoplethic maps should generally be used to map standardized data (e.g., population density instead of raw population counts) (Figure 9.12). They are ideal to use instead of a choropleth map when mapping continuous and smooth data. They naturally represent a smooth transition between data values compared to the abrupt enumeration unit changes found on choropleth maps.

Isarithmic Map Design

Several best practices have evolved in print cartography regarding isarithmic map design that still hold true. First, choose an appropriate isoline interval. The interval between values will determine how many lines fill your map. Too much detail in isarithmic mapping makes the map difficult to read and results in someone missing broader data trends for area specifics. Second, label only every fourth or fifth isoline. Isolines that are labeled are called index lines and often have a slightly thicker line width. Third, fill in the areas between isolines with appropriate color schemes to facilitate a clear under-standing of which direction data are going—up or down.

Benefits and Drawbacks of Isarithmic Maps

Isarithmic maps, like all others, have their pros and cons. Generally, they are easy for people to read. This is particularly true when a color scheme is added to the lines to symbolize increasing or decreasing values. Unfortunately, they can be a bit more difficult to create using standard Web technologies. It is therefore often easiest to create a map in a GIS or statistics program and then import the image into your Web map. Another issue is designing smooth and fluid isolines. When one uses a statistical package or GIS to create isolines, they often end up appearing a bit jagged. Isolines are best when they look natural and fluid. Touching up rough-looking lines in a graphics software package or GIS before exporting them for inclusion in your Web map is strongly recommended.

Flow Maps

Flow maps are one of the most dynamic and useful thematic representations at a Web cartographer's disposal. Due to a variety of factors, they are also the least likely to be used. Flow maps are used to represent data that move. Most commonly, this is done by drawing a line with an arrowhead showing which direction a movement is occurring (e.g., exports out of a country). When bidirectional movement is occurring either two or sometimes no arrowheads are shown (e.g., total highway traffic between two cities).

In addition to movement direction, a vast array of visual variables can be used to highlight quantitative and qualitative data, sometimes in the same map. Most commonly, line thickness is set to be proportional to a data value (see Figure 9.13)—proportional flow maps. The values can also be classed to make graduated flow maps as well (i.e., predetermined line thicknesses representing a range of values).

Lines can also be coded using hue, value, and saturation (Figure 9.13). Hue might be used to differentiate among ethnic groups on a global migrations map. Color value and saturation can be used to supplement line thickness or add another quantitative variable on top of line thickness. For example, perhaps line thickness on the flow map represents total number of goods being shipped, whereas the color value of lines represent what percentage of these goods are considered high-value ones.

Finally, instead of using line thickness, one can create a heat flow map. Heat flow maps are actually a combination of dot and heat maps using lines instead of points. Each line represents a predetermined value (e.g., 20 illegal shipments of paprika). If one country, let us just say Hungary, has 500 illegal shipments leaving it every year, there would be 25 lines of equal width leaving Hungary's borders. Depending on which direction these lines were all heading, and let us presume they are mostly heading toward the Benelux countries, they will start to overlap. As the lines overlap, the density of lines increases, and the color of the lines changes from blue to green to red.

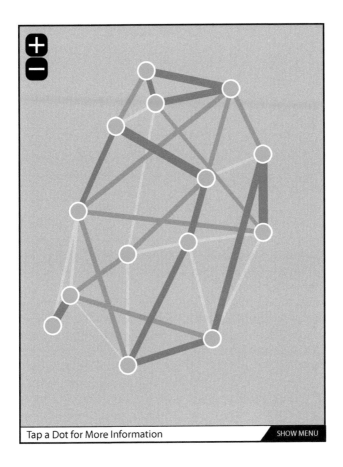

Tap a Dot for More Information   SHOW MENU

FIGURE 9.13
Flow map.

Benefits and Drawbacks of Flow Maps

Interactive Web mapping provides another incentive to start creating flow maps: animation. Flow maps are no longer restricted to showing movement merely via lines and arrows; now, these lines and arrows can be animated to show movement over time. Animation and flow maps go together like peanut butter and jelly or, for my European audience, chocolate hazelnut spread and baguettes. Because Web mapping allows for animation, one might think that flow maps would become ubiquitous; yet, they remain relatively rare.

Designing flow maps still requires a lot of grunt work on the part of cartographers. Even in print, flow maps are sparse compared to many of the other thematic representations. Few algorithms, plug-ins, or extensions exist to create flow maps in GIS software. Those that do exist often result in less-than-stellar outputs—typically jagged, robotic-looking straight lines that lack the aesthetic pleasantness of Bézier curves. Many flow maps are still drawn by hand (i.e., computer mouse) using graphics programs. These data can then be exported as vector art and imported into a Web map. However, when it comes to doing all of that work versus just creating a choropleth map of Europe showing each country's percentage of illegally imported Hungarian paprika, most people will likely settle for the latter.

Cartograms

At the start of this section, I defined thematic maps as spatial data mapped over a base map. This is by and large the simplest definition of thematic maps; yet, it omits one style of map: cartograms. Cartograms differ from other thematic maps. Instead of visualizing data over a base map, cartograms represent data by directly distorting the shape and size of enumeration units on the base map. The size of enumeration units is changed based on a common data value and made proportional to one another. Cartograms are most effective at mapping raw data, but they can also be used to map percentages that, when combined, equal 100.

There are two main categories of cartograms: contiguous and noncontiguous. Both distort the area of the enumeration unit. Contiguous cartograms maintain the topology of the enumeration units. Maintaining contiguity among places means they should be relatively easy to find. In reality, however, the shapes of enumeration units are so distorted and stretched that it becomes extremely difficult to find the exact place you are looking for. Also, because enumeration units are so warped, it can also be difficult to determine the difference in values between several places unless your brain is really good at visual calculus. What contiguous cartograms do well, however, is highlight the proportion of one enumeration unit's value compared to all others (see Figure 9.14).

Traditionally, noncontiguous cartograms resize all of the enumeration units proportionately as well. However, each enumeration unit is separated from others so that its shape is not distorted. This makes it easier to find the enumeration unit you are looking for. The problem is, it is often more difficult to determine the percentage of the total value that an enumeration represents because all of the symbols are spread out. An alternative approach to constructing noncontiguous cartograms was proposed by Daniel Dorling (1993) and has become widespread. Dorling developed a method of creating cartograms out of geometric instead of real enumeration unit shapes. Once the enumeration units are sized appropriately, they are separated so that no overlap exists. The topology of the enumeration units is retained as much as possible, resulting in an easier to decipher cartogram (see Figure 9.14).

Benefits and Drawbacks of Cartograms

Cartograms are extremely useful due to their novelty. They often visually distort data so wildly that they are exciting to look at. This is both a blessing and a curse. You generally want your audience to be excited about your map

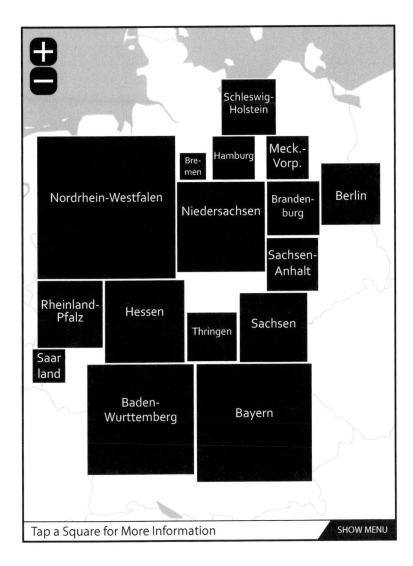

FIGURE 9.14
A noncontiguous cartogram using geometric shapes (styled after Dorling).

as this helps you achieve your communication goals. However, cartograms are also difficult for some people to read and make sense of. Adding information windows to your cartogram can help overcome some of these limitations.

Multivariate Maps

Multivariate maps are maps that show more than one theme at a time. Web mapping has made this incredibly easy to do as numerous datasets can be mapped all at once, and map users can turn different representations (or layers) on and off at their whim. Here, I review five common types of multivariate representation: thematic map combinations, colored dot maps, chart maps, Chernoff faces, and bivariate maps.

Thematic Combinations

Thematic combinations are simply the layering of multiple thematic representations on top of one another (see Figure 9.15). For example, you may have

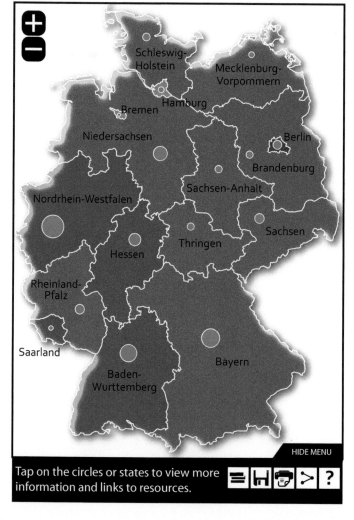

FIGURE 9.15

German population density and population mapped using choropleth and proportional circle techniques.

a choropleth map with a proportional symbol map on top of it. Thus, map users could visually compare how two spatial variables correlate. If the variables being mapped are relevant to each other (e.g., population density and raw population numbers), such maps can be enlightening and are typically quite easy for most map users to interpret.

Colored Dots

Dot maps can show more than just the distribution of one variable. By changing the colors or shapes of dots, dots can be used to show the distribution of many variables. This is a highly effective and intuitive way of showing multiple point data at once. For example, one might show Tweet data in Washington, D.C., on an election night. You could represent Tweets referring to the Democratic Party in blue, those referring to the Republican Party in red, and those referring to other political parties in black. The only problem with multicolor dot maps is that, depending on the resolution of your data and the scale of your map, the dots may start to overlap and become indecipherable.

Chart Maps

Chart maps are those that use charts to represent multiple data values at once over enumeration units. This is commonly done using a pie chart. Even better, the charts themselves can be sized as proportional symbols, adding yet another data layer to one's map. For example, one might design a proportional circle map representing the population of German provinces. Each proportional circle could then be made into a pie chart showing another variable (e.g., the gender distribution of the population).

Bar charts can also be used as a method of representation and frequently are found in print maps. However, using bar charts as a thematic symbol in Web maps is rare. Due to their less-compact vertical extents, proportional bar charts are often more distracting and difficult to read than proportional pie charts.

Chernoff Faces

Chernoff faces can represent over 10 different pieces of data at one time using one symbol that looks like a face. They were first proposed by Herman Chernoff in a 1973 statistics paper. By combining all of these independent facial features into a face symbol, it was possible to show how a multitude of variables correlate with one another (see Figure 9.16). When you place these faces over a map, you can see spatial trends. While Chernoff faces are really neat, they are not really all that useful for conveying detailed information due to the fact that they are so complex. Still, they have aesthetic and rhetorical usefulness in a variety of circumstances, and they are becoming easier to

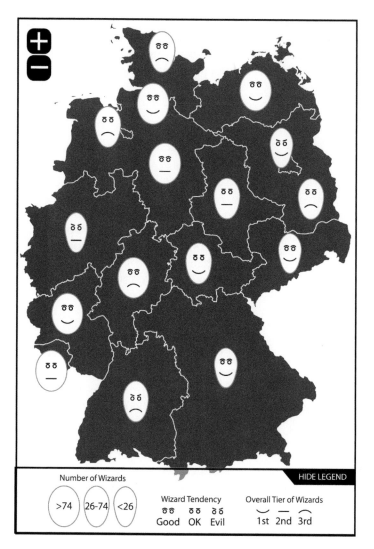

FIGURE 9.16
An example of a Chernoff face map.

create thanks to new software tools (see Yau's *Visualize This* [2011] for more information on creating your own Chernoff faces with free software).

There are several things one can do to make Chernoff faces more intuitive and readable. First, you should always generalize data to the ordinal level (e.g., low, middle, high). It is much better to have only three or four data variations per facial expression. Also, the legend should clearly show how each attribute is constructed and what each facial feature means. Finally, you could have an info window appear or data in a side panel when someone selects a face to provide exact statistics for its features.

One thing that should be mentioned is never to use color as a variable when designing Chernoff faces. Use the same color for each face. Chernoff faces have been used to map demographic data using a white-to-brown color scheme. As you can imagine, the scheme had serious racial connotations and incensed many people. Other choices in facial features can also be seen as racial stereotypes and should be carefully thought through so that you do not offend your audience.

Bivariate Choropleth Maps

Bivariate choropleth maps use variations in hue, value, and saturation to map two dimensions of data at once (see Figure 9.17). When well designed, these maps are aesthetically stunning. However, they are almost always difficult for someone to read accurately. A bivariate map user typically must recall at least nine different classes while looking at a map. Unless your audience is willing to spend a considerable amount of time looking at your thematic map, I recommend avoiding this visualization technique in Web mapping.

The Role of Interactivity in Thematic Representations

Since the 1990s, cartographers have conducted a variety of research projects regarding the benefits of map user interactivity with thematic representations.

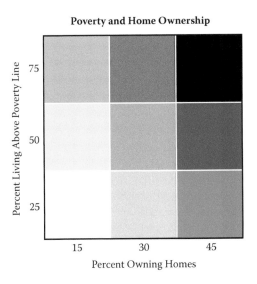

FIGURE 9.17
A bivariate map legend.

It turns out that interactivity can be a mixed blessing. Too little, and map users become frustrated. Too much, and they become distracted from the message of the map.

There are some who believe that more human-map interactivity will generally help users make better sense of the data. The argument is libertarian in nature; people know what is best for them and will manipulate the data so they can more easily interpret them. Allowing people to interact with a map can promote data exploration and increase interest in what is being shown. For example, many times map users are given tools to reclassify the data, change the color schemes or shapes of thematic symbols, and turn different representations on and off. By being able to change how data are represented, users have the potential to see information in a way they otherwise would not. If detailed interpretation of the data underlies the purpose of your map, then it is probably a good idea to include more interactivity and allow map users to manipulate how the thematic maps present the data.

In some cases, however, data exploration is the last thing you want your map users to do. If the purpose of your map is to communicate a clear message, you should strive to limit the impact a map user can have on your classification and aesthetic design decisions. That way you, the mapmaker, are in charge of the production and the communiqué.

As discussed throughout this book, a bit of map user control over a Web map is generally a benefit. Map users like to feel like they are empowered. However, too much interactivity can allow a map user to completely miss the point of your map and become easily distracted. Unfortunately, there is no right or wrong answer as both philosophies are right some of the time. The utmost responsibility of a Web cartographer is to design maps that achieve their communication goals in an aesthetically appealing manner that is also palatable to the intended audience.

Conclusion

Designing thematic maps is one of the most exciting things one can do in cartography. Taking data and refashioning it into information about a particular real-world phenomenon is the first stage of thematic mapping. The next step is to find a suitable method of representation that respects the data and phenomenon's spatial characteristics. The third stage is stylizing this representation in a manner that will be readily interpreted by an audience. Finally, you must decide just how much control and manipulative power map users should have over your final representation.

One chapter is far too short to do justice to thematic cartography. In no way does this represent a comprehensive list or discussion of thematic representation. There are many types of thematic maps that I did not have space to discuss here. Under "Further Reading" I provide a variety of books, blogs, and journal articles you can look at for more information on different thematic visualizations.

Key Concepts

- Spatial data are not the same as spatial phenomena.
- Thematic maps should be created using equal-area projections.
- There are many types of thematic representation beyond the choropleth and dot map.
- Different thematic maps are better for displaying different types of data. It is important to know when each is appropriate.
- Allowing map users to manipulate thematic data and how they are presented can be either beneficial or disadvantageous. One must therefore think about what will happen to the map's message by allowing Web map users more or less control.

Further Reading and Resources

Resource

Flowing Data Web site. http://www.flowingdata.com. Flowing Data is a Web site by Nathan Yau. He highlights interesting visualizations. He also posts a variety of visualization tutorials (http://flowingdata.com/category/tutorials). I recommend the one on creating Chernoff faces with free software.

Further Reading

Chernoff, H. (1973). The use of faces to represent points in k-dimensional space graphically. *Journal of American Stastistical Association*, *68*(342), 361–368. Retrieved from http://amstat.tandfonline.com/doi/abs/10.1080/01621459.1973.10482434

Dooley, M., & Muehlenhaus, I. (2010). Mapping UFO sighting data: pitfalls and possibilities. NACIS Annual Conference, October, St. Petersburg, Florida. Retrieved from http://www.nacis.org/documents_upload/NACIS2010.pdf

Dorling, D. (1993). Map design for census mapping. *The Cartographic Journal* 30, no. 2: 167–183.

Gigerenzer, G. (2007). *Gut feelings: the intelligence of the unconscious* (p. 280). New York: Penguin Group.

Gladwell, M. (2005). *Blink: the power of thinking without thinking* (p. 296). New York: Back Bay Books.

MacEachren, A. M. (1995). *How maps work: representation, visualization, and design* (p. xiii, 513 pp.). New York: Guilford Press.

Yau, N. (2011). *Visualize this: the flowing data guide to design, visualization, and statistics.* New York: Wiley.

10

Animation

Introduction

Static data visualizations aside, one of the most fundamental shifts that digital cartography has brought about is the ability to animate one's maps. What is map animation? Essentially, it is the use of different motion techniques (largely devised by the film and cartoon industry over the past 100 years) to display data change on a map.

We have grown so accustomed to animated maps within the past decade that it is easy to forget that animated mapping has been a long time in the making. (For fascinating reading on the history and evolution of animated cartography, see Harrower, 2004.) Animated maps began appearing as cartoons and newsreels beginning in the 1930s. Within 30 years, there was already an academic publication, *Animated Cartography* (Thrower, 1959). Animated cartography remained a bit of an oddity that was experimented with off and on over the years, most notably by Waldo Tobler, who created the first computer map animation (Tobler, 1970).

Though map animations became more common and easy to view with the rise of the household personal computer (PC), animated cartography was never really that marketable until the invention of the Web. Certain hardware requirements were necessary for the technique to become ubiquitous. Namely, one needed a better method of map distribution than discrete media (e.g., videocassettes, computer disks, CD-ROMs, or DVDs). The mass adoption of the Internet beginning in 1994 facilitated the rise of animated maps. Suddenly, people could easily access animated maps without needing to order and plug in discrete media. Moreover, they were often afforded some control over how these maps played back.

This chapter provides fundamental information about the nature of animation that you should take into consideration when designing animated maps. Although animated maps are becoming easier to make and are often popular among audiences, there are many things that need to be considered before jumping in and designing one's own.

To Tween or Not to Tween?[*]

Methods of animation can be broken down into two broad types: stop frame and tweening. Each has its own unique advantages, and Web mappers should determine which type of animation is most appropriate to use before they begin designing their maps.

Stop-Frame Animation

Stop-frame animation (also known as "stop motion") is one of the oldest types of animation in existence. It is the process whereby individual frames (originally film but now digital) are designed one by one. Originally, this was done by taking a picture of something and then slightly manipulating the object (e.g., moving it) before taking another picture. After many pictures and movements are made, one can view the pictures quickly in the order they were taken to view a stop-frame film. This technique of animation often results in a lot of manual work when dealing with vector graphics, and the animations can sometimes come off as quite jerky depending on the rate of change (discussed in the next section). Nevertheless, stop motion can be a very effective aesthetic technique. Moreover, for time-series raster data, it is typically the only animation option available. For example, you may show the same map image from a particular year for 10 seconds, and then in the next frame, the map will quickly switch to a new data snapshot from a different year.

Tweening

Tweening is used to create smoother animations than the stop-frame method. The term *tweening* comes from the word *betweening*. Tweening is the process of interpolating movement between key frames (i.e., frames on the timeline with a defined image). For example, in Figure 10.1 the first frame is a key frame, and there is another key frame in frame 9. Tweening uses image interpolation to fill in the seven frames found between the two key frames.

Tweening is good for showing continuous temporal data. This is data that do not have temporal gaps (e.g., an animation of all commercial flights originating in Europe over the course of a day). By its very nature, tweening implies smooth, continuous change. It should generally be avoided for showing information with large temporal gaps. For example, tweening would be great for showing an animation of an invading army conquering large swaths of land. These are continuous and smooth data, showing land acquisition and acquiescence over time. However, tweening is not ideal for

[*] Note: "To Tween or Not to Tween?" was borrowed from the title of a paper presented by Dr. Matt Dooley at the North American Cartographic Information Society Annual Meeting in 2011 in Madison, Wisconsin. All credit to him for the witty title.

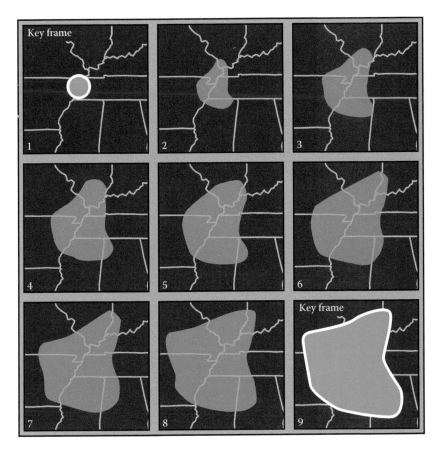

FIGURE 10.1

Tweening is the interpolation of frames found between two key frames. In this image, frame 1 is a key frame with an initial area. Frame 9 is also a key frame. Frames 2–8 represent an interpolation, a tween, between frame 1 and frame 9. This results in smooth and continuous animation.

showing the number of gold medals won by different countries in each Olympics. There are 4-year gaps between each Olympics, and the data from one Olympics have no direct correlation to the data from other Olympics (different athletes, competitions, and circumstances).

The Visual Variables of Animation

As already mentioned, map animation has been studied for well over 50 years. However, it was not until Bertin's work that people began thinking in depth about how visual variables might apply to animated maps. By the early 1990s, it was becoming quite apparent that several additional variables of representation were at play in animations.

DiBiase's Visual Variables for Animation

David DiBiase, MacEachren, Krygier, and Reeves (1992) identified and labeled three new visual variables dealing specifically with animation: duration, rate of change, and order.

Duration

Duration is how long a particular map frame (i.e., a still image) in an animation is shown before the next frame in the animation is presented. Duration can also apply to groups of frames showing the same exact thing—typically referred to as "scenes." For example, assume you are animating state change in populations using decennial census data in a choropleth map. The data only change every 10 years. You will want to let map users view each decade's population information for at least several seconds before switching to the next decade's dataset. Therefore, if your animation is set to play 24 frames per second (fps), you might decide to show the same 1990 second census data for 96 frames, resulting in a 4-second duration.

This is merely duration in its simplest form. As a visual variable, duration is ideally suited to represent quantitative differences (Figure 10.2). It can be used much like a proportional symbol on a static map. Variation in the length of duration corresponds to the quantity of a variable (e.g., time, population, etc.). For example, if you were animating countries that came in first in the gold medal count for each Olympics since 1896, you could change the duration of each year's scene depending on how many gold medals the first-place country won that year compared to how many were won by first-place countries in other years.

Rate of Change

Rate of change represents how quickly an image is morphed into or switches to the next attribute. Specifically, rate of change is the magnitude of an attribute (e.g., population size) divided by the duration (i.e., how long it is shown). It is represented using m/d (magnitude of change divided by duration).

Duration is already understood; magnitude is the new concept. Magnitude of change is how much the animated image being represented changes from frame to frame—a lot or a little. One can have a longer duration (e.g., 72 frames at 12 fps) and a small magnitude of change (resulting in a smoother animation). On the other hand, one could have the same duration but a large magnitude of change (potentially resulting in a less-smooth animation but faster display of the information). Figure 10.3 highlights the influence that

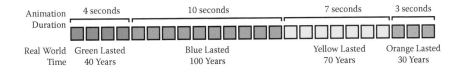

FIGURE 10.2

Duration represents the manipulation of how long something is shown based on an attribute or value. In this example, we animate fictitious empires (represented by color). Each ruled for a different length of time. The duration of each empire's representation is proportional to how long it ruled in the real world.

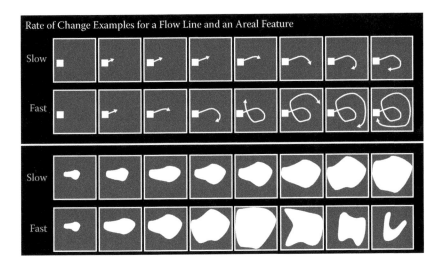

FIGURE 10.3

Two examples of the impact of different rates of change on animations. On the top, two rates of change for a flow line emanating from a point (e.g., my dog wandering in the backyard). On the bottom, two rates of change for a real unit changing in size and shape (e.g., the fate of an empire in a video game). Faster rates of change show more data in the same amount of time.

different rates of change would have on two animations; one shows a flow arrow moving and the other an areal feature changing in shape and size.

Rate of change can often be used for dramatic effect. You might start an animation off slowly, showing data at a relative moderate pace. As the animation continues, however, you might speed up the rate of change, resulting in a faster presentation of data. This effect is often used in persuasive maps. One might show the spread of Wal-Mart stores beginning slowly and then speed up the rate of change as more and more stores are founded. It can create a tsunami of data toward the end of an animation, which can be rhetorically quite powerful. Less surreptitiously, you may want to allow your users to manipulate the rate of change interactively so that if an animation is moving too fast for them, or if they want to analyze a subset of data in more detail, they can.

Order

In addition to duration and rate of change, the order of how things are presented can be used as a visual variable. Rather than present data in the order they actually happened, you can reorganize data presentation by value. For example, the countries that have earned more than 50 gold medals in the history of the modern Olympics might be shown in descending order based on how many gold medals their athletes have won, beginning with those countries that have earned the most gold medals (see Figure 10.4).

Additional Visual Variables of Map Animation

Others have added to DiBiase's original three visual variables. MacEachren (1995) concluded that three more variables exist: display date, frequency, and synchronization.

Display Date

Display date is a simple visual variable. It represents when a place or variable is first shown within an animation (see Figure 10.5). For example, the United States won the most gold medals in the first Olympics in 1896. Therefore, in a

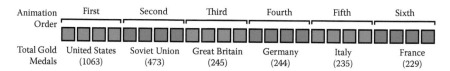

FIGURE 10.4
Order is the rearranging of an animation so that spatial data do not show up chronologically but instead are based on another attribute value. In this example, countries are animated on a world map depending on how many gold medals they have received throughout the history of the modern Olympics. The countries that have earned the most medals are shown in descending order for the same duration.

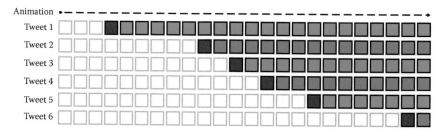

FIGURE 10.5
Example of display date. Different tweets arrive at different points in the timeline. A tweet's arrival on the screen *is* its animation, so to speak. With hundreds of tweets popping up on the screen over the course of the timeline, an illusion of continuous animation is created.

chronological animation of winners of the gold medal count, its display date is immediate. France's display date is 1900, when it won the most medals, so its display date would occur slightly later than the U.S. date in the animation. Obviously, display date tells someone when something arrives on a scene. It is particularly useful for creating temporal animations.

Frequency

Frequency represents the amount and regularity of an attribute occurring within the object being animated (Figure 10.6). An example might be the use of background shading (from sunrise to high noon to sunset and then through the night) to symbolize different periods of each day in an animation of a boat circumnavigating the world. This shading would give a clue regarding how far the boat had traveled at different points during a day, while the overall animation would show the route and duration of the trip around the world in its entirety.

Frequency is often implemented when subsets of data need to be highlighted within an animation. For example, one might want to animate how regularly chickens have been stolen in Graz, Austria, over the course of the past 6 years (Figure 10.6). Watching the animation, map users would be able to ascertain how often in a given year chickens were stolen by counting the frequency of animated dots (each representing a chicken) that appear throughout the that year.

Synchronization

Synchronization refers to two or more animated datasets (Figure 10.7). This visual variable can be used to highlight temporal correlations and relationships between two different types of data. For example, the number of fatalities in a civil war might be animated concurrently with an animation showing the amount of foreign aid given to rebel and government forces.

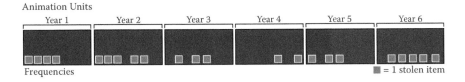

FIGURE 10.6

Frequencies provide the map user information about how many events occur within a given unit of time. In this example, a year might be represented by 10 s. A stolen item (e.g., a chicken) is represented by a red box. During the animation, people can determine how frequently chickens were stolen per year and when they were stolen within a year. Stolen chickens might be represented via a dot flashing on the map or perhaps via sound—the sound of a chicken clucking. The map user can count the frequency of this while keeping tabs on the timeline.

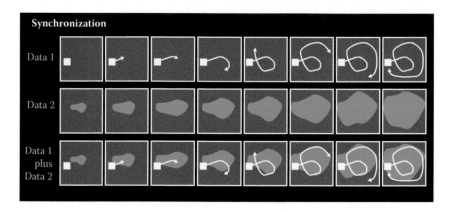

FIGURE 10.7

Synchronization is the process of highlighting how two spatial elements correlate both spatially and temporally. In this example, both a path feature and an areal feature are animated using the same rate of change. When they are overlaid, one can see whether there is a temporal-spatial correlation between them.

The synchronized animation might highlight a correlation between foreign intervention and war dead. This visual variable is extremely useful for displaying relationships between temporal datasets.

Types of Map Animation

Knowing the visual variables of animation is important so that you can actually think about how best to represent your data. In this section, I categorize different types and styles of map animation that you may find useful to incorporate into your maps, and I discuss when it is appropriate to use different animation methods. As will be shown, by far the most common use of map animation is temporal (i.e., showing change over time in a chronological order); however, this is not the only way animation can be used to spruce up your maps.

Designing Temporal Animations

Most of us view an animated map displaying temporal data almost every single day (i.e., during a weather forecast or on a weather Web site). For many of us, animations are synonymous with temporal representations. The usefulness of temporal animation for representing data is potentially enormous.

I use the word *potentially* on purpose. Although showing change over time via animation seems like a no-brainer, research has shown that people do not always interpret animated data accurately. Temporal animations are often better at conveying broad patterns of change in a dataset than they are at showing differences between particular places. In fact, the research (please see the "Further Reading and Online Resources" section) concerning the usefulness of temporal animations and how well humans perceive them is so muddied, that I recommend only using animations to portray broad changes at a small scale. Detailed changes should only be animated at a large scale. Numerous issues may affect how well people interpret your animated map, including "change blindness," in which obvious events are completely missed due to a map user focusing on something else. Although many of us may have trouble interpreting animated maps, it is conventional wisdom that people love how animations look. To help keep your map users happy and loving your animations, keep the three following truisms in mind when designing animated maps.

Keep the Animations Short

First, keep animations short. Anything too long will overwhelm a user's short-term memory, and data retention will soon be nullified. Most thematic map animations should not last more than 30 seconds to a minute. This is particularly true if the temporal data you are mapping are not narrative in nature but rather show changes in a dataset over time. Narrative maps, on the other hand, can be longer because they typically are not presenting as much data to a map user and allow for pauses to reflect and read about events.

Simplify the Data

Second, simplify the data. Human perception and cognition are limited. Some of these limitations are imposed due to evolutionary factors; many, however, are simply due to cognitive overload (i.e., information overstimulation). When cognitive load becomes too heavy, little is meaningfully absorbed into memory.

Animation can result in visual overstimulation. To glean information from an animation, we can really only focus on a few things. Frequently, data in a Web map animation cover the entire mapped area, which means the map user is expected to perceive a wide array of animation, not just focus on a concentrated part of the screen. Moreover, the animated representation might be displaying quantitative values, so the map user is expected to take note of a wide array of different data values (and their locations) while perceiving changes in these data values across the entire map.

Thus, it is best not to use temporal animation to communicate a message that is too complex for anyone to comprehend—no matter how cool it looks. One thing to keep in mind is that humans can only focus their vision on three to four things at any single moment (Ware, 2008). Everything else we think we see is largely based on the short-term memory of previous eye movements. We move our eyes around constantly to take in information, but we quickly forget what we just saw. In the end, by moving our eyes around a lot on an animated Web map, we can ascertain general trends and patterns, but typically, we cannot remember specific details about the data we just viewed.

Give the Map User Some Control

Third, people like to feel like they have control over animations and their playback, so always give the map user some control over the animation. If you provide no interaction with an animated map, map users will often feel powerless and potentially frustrated when they want to rewind or pause an animation and cannot. Providing a timeline, a play button, a pause button, and at least a simple rewind button is generally a good idea. YouTube, Hulu, and Netflix, among others, have made these simple interactive buttons standard for almost every animation we view. Therefore, it is best to include them in your animated Web map as well.

Temporal Legends

Interactive legends were covered in Chapter 3. However, temporal legends are different. Such legends are both map elements and integral parts of animated visualizations. Making sense of an animation is nearly impossible without them. Adding the fourth, temporal, dimension to your representation is useless from an information standpoint if people have no idea when the data they are viewing occurred or how quickly the animation is moving. These legends can take many guises, and although there are three favorite techniques of representing time in a legend form, one should not shy away from developing new ones. There simply is no accepted standard when it comes to animated legend formats yet. Perhaps you will invent one that will be widely adopted.

Temporal Legend Styles

The first method of showing time on your map is simply to add a clock (Figure 10.8C). The benefit of adding a clock or calendar-type component is that the user can quickly look at it to see when something is occurring.

A second legend type, and perhaps even more common these days, is a temporal bar scale (Figure 10.8B). A temporal bar scale is similar to the timelines found in nearly every video viewed on the Internet. Typically, these are found to the right of a play button at the bottom or top of the video. The best ones are interactive. If one taps on any part of the timeline, the movie will jump to that part of the timeline and continue playing. Temporal bar scales work the same way as these movie timelines. They can be designed to be interactive or not. Generally, map users like being able to pick a time spot in the animation. The main point of a temporal bar scale is that, as the animation plays, the temporal bar scale shows how much time has elapsed, how much time remains, and where the map user is on the timeline at any given instant.

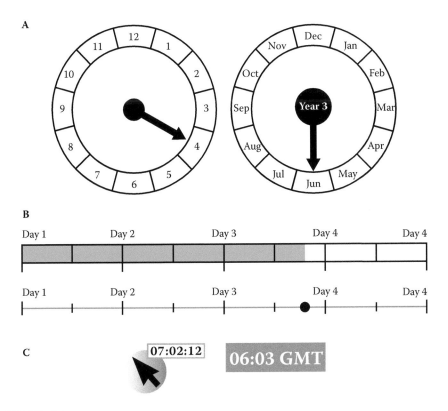

FIGURE 10.8
Different temporal legend examples: (A) Two examples of different temporal circles; (B) two examples of different temporal bar scales; and (C) two examples of a clock, the second of which is activated by right or left clicking anywhere on the map.

A third type of animation legend is a temporal circle (Figure 10.8A). Temporal circles are similar to a temporal bar scale, but they are more compact. The benefit of using circles is that if you are mapping something that is cyclical (e.g., occurring over 24 h or over the course of four seasons), it allows you to represent time far more compactly. Also, a circular legend infers that a process is cyclical. For example, one might represent time moving around the outside of the circle. The circle itself might be divided into 12 months. In the middle of the circle, the year might be placed. Every time the animation plays through the 12 months, the year would change to let the map user know a new year has begun.

Temporal Legend Enhancements

The downside of temporal legends, regardless of type, is that map viewers will likely need to switch their attention from the thematic data represented to determine the time at which a phenomenon occurred. This need to switch-task, as it is often called, will inevitably result in less-than-efficient information gain and retention. Each visual referral to the temporal legend will result in a visual and cognitive distraction that will result in the map user missing information and will potentially muddle your communication goal. One can add several enhancements to temporal legends to make them even more effective.

Always place the temporal legend as near as possible, if not partially overlapping, the mapped area. Obviously, you do not want your legend to obscure important data, so take care not to infringe on the spatial information being presented. Another technique is to have a little clock pop up when someone right clicks with a mouse. As has been discussed previously, though, there are serious limitations to assuming someone will interact with your map using a mouse. A third option is actually to have audible or verbal confirmation of particular index times. For example, if you are showing change over the course of a day, you could make a small noise play at the top of every hour. Alternatively, you could have a voice verbalize what time it is. This way, a user will not have to look at the clock to gauge what time it is.

Placement of a temporal legend is also important. Temporal legends are often placed at the bottom of a Web map. This makes some sense for temporal bar scales, as people tend to expect temporal bars to be found at the bottom of whatever online film or animation is playing. However, this may not always be the most intuitive place for these legends. Placing the temporal bar at the top of the map where it is more prominent and closer to the visual center of the mapped area might make more sense; it is a shorter distance from the mapped area to the legend, therefore potentially less of a nuisance for the map viewer. Moreover, if a temporal legend (regardless of type) fits well with other interactive map elements (e.g., thematic legends, zoom bar, or info box), it should probably accompany them so the map user does not have to move his or her eyes all over the map to find what the user is looking for. Finally, whenever possible, make sure the temporal legend is not overly gaudy or distracting. Given all of these things to think about, conducting some user testing on temporal legends before putting them online is highly recommended.

Other Types of Map Animation

Temporal map animations dominate the discussion, but numerous types of animation are used every time you use a service like Google Maps, Bing Maps, or Google Earth. Three of the most common and important types of animation in addition to temporal are zoom, fly-through, and path animations.

Zoom Animations

Zoom animations are those that occur when a map's cartographic scale is changed in the Web map. (Please see Chapter 3 for a complete rundown on zoom features.) Zoom animations can be preprogrammed within a Web map; this is often done with narrative or storytelling-style maps. For example, a map showing the island-hopping campaign of U.S. forces in the Pacific during World War II may start at a small, hemispheric scale to show the battles in a broader context before zooming in to show large-scale maps of different islands. More commonly these days, the map user is provided the option to change a map scale interactively. This is done almost every time someone looks up directions on a map service provider site. Zoom animations can also be useful on thematic Web maps that have very small enumeration units that are difficult to see when viewing all of the data at once.

How zoom animations function varies depending on the nature of the data being displayed. Many, if not a majority, of Web maps these days use raster tiles for most of their data representation. You can generally figure out which maps are using raster tiles when you zoom in on the map. As you start the zooming process everything gets bigger, and pixelated, before refreshing to tiles showing more detail at the newly selected scale. Increasingly, vector data are being used in Web maps (e.g., Scalable Vector Graphics [SVG], Web Graphics Library [WebGL]). When you zoom in on this type of map, the data will become larger but will not pixelate.

Fly-Through Animations

Fly-through (or increasingly drive-through) animations are those that swoop over or across a landscape. The goal of these is to give the impression of flying over a scene. It is known that many people prefer looking at objects from a bird's-eye view (i.e., slightly oblique, not completely top down, as shown in Figure 10.9) because this is how we typically envision objects in our mind (Weinschenk, 2011). Thus, it stands to reason that people will also like looking at mapped data from this angle. This is one reason many of the major map providers (e.g., Microsoft Bing, Google, and Apple Maps) are spending a lot of money on interactive bird's-eye views of cityscapes.

Fly-through animations are often useful in narrative maps or to provide an overview of terrain in an area. These types of animations, though, raise several issues that other types of animation do not. The first problem is that people can quickly become disoriented. Thus, any time you include a fly-through animation, it is good practice to have an animated, two-dimensional locator map as well so people can refer to where they are in the grand scheme of things. Second, labeling in 3D fly-through animations can be very difficult. If labels due not realign themselves with the movement of the map user, they may end up being illegible or backwards. On the other hand, if the labels rotate, they may distract from the data you are trying to show.

FIGURE 10.9

An example of a fly-through animation using Google Earth. (Copyright 2013 TerraMetrics, 2013; Google, 2013, CNES/SPOT.)

Path Animations

Path animation is useful for giving directions and for guiding a reader along in a narrative map (see Figure 10.10). Path animation highlights a route from start to finish. The animation is essentially meant to highlight the path for a user. It may be a temporal animation in a sense, but not always. For example, instead of highlighting the amount of time a route takes, the animation may instead emphasize landmarks, key intersections, or other focal points of a narrative. An example of this would be a map highlighting John F. Kennedy's route on the day of his assassination in 1963 in Dallas, Texas. The animation of the route may stop at various points to highlight certain pictures or information (e.g., a certain book depository or grassy knoll). The path animation would be chronological, but the purpose of the animation would be to highlight where things are in relation to one another spatially and in an ordinal sense, rather than in a temporally quantifiable manner.

Summary

Animation is a great tool for catching and focusing people's attention on specific parts of a map. It can be used to enhance your Web maps or as the primary data of the map itself. Animation can also be used to highlight

FIGURE 10.10

An excellent example of a path animation of "The Story of the Donner-Reed Party" by Amy Lippus. This map shows the travels and travails of this group of people as they migrated westward. These types of animations are stellar for representing stories; they allow for highly engaging narrative mapping. (Used with permission. Available at http://www.csuchico.edu/geop/announcements/CGS%202012%20-%20Amy.swf.)

change over time, create engaging narrative maps, and allow people to view data from unique angles and perspectives. That being said, it is also a risky tool. Animation, when not used in moderation, can become more of a distraction than a benefit. It can quickly overwhelm map users and blot out your map's communication goals behind a wall of visual noise. When designed well, it will make your Web map extremely appealing and informative.

Key Concepts

- Additional visual variables for animations can be used to imbue your animation with quantitative and qualitative values. These include duration, rate of change, order, display date, frequency, and synchronization.

- Animations can show more than temporal change. They can be used to show location (e.g., blinking symbols to highlight an element's display date), illustrate how data are spatially distributed across a landscape, and emphasize the area something consumes at a particular time.
- Keep your map animations short. The more complex the data, the shorter an animation should be. Humans simply cannot maintain an attention span long enough to retain lots of information from an animation.
- Simplify the data used in animations. Ordinal-level data (rankings) are often more easily interpreted than metric (measured) data.
- Always include a temporal legend of some sort.
- Always give the user some control over an animation: play, pause, forward, back, or the ability to click on and fast forward or rewind to particular key frames.
- Fly-through animations can be quite exhilarating. On the other hand, people rarely have the cognitive capability to truly understand direction or distance in such map scenarios. Include a locator map when using these.

Further Reading and Online Resources

Online Resources

Animaps. http://www.animaps.com
FlowingData. http://www.flowingdata.com

Example Animations

Boice, J., Bycoffe, A., & Scheinkman, A. (2013, March 22). 98 days since Newton. *Huffpost Politics*. http://data.huffingtonpost.com/2013/03/gun-deaths
Earth Observatory. Earth at night animation. http://earthobservatory.nasa.gov/NaturalHazards/view.php?id=79803
Lippus, A. (2012). A Californian tragedy: The story of the Donner-Reed party. http://www.csuchico.edu/geop/announcements/CGS%202012%20-%20Amy.swf
Swenson, D. (n.d.). Flash flood. *Times-Picayune*. Katrina Map: http://www.nola.com/katrina/graphics/flashflood.swf

Further Reading

Andrienko, N. V, Andrienko, G. L., & Gatalsky, P. (2005). Impact of data and task characteristics on design of spatio-temporal data visualization tools. In J. A. Dykes, A. M. MacEachren, & M.-J. Kraak (Eds.), *Exploring geovisualization* (pp. 201–222). Oxford, UK: Elsevier.

Battersby, S., & Goldsberry, K. (2010). Considerations in design of transition behaviors for dynamic thematic maps. *Cartographic Perspectives, 65,* 16–32, 67–610.

DiBiase, D., MacEachren, A. M., Krygier, J., & Reeves, C. (1992). Animation and the role of map design in scientific visualization. *Cartography and Geographic Information Systems, 19*(4), 201–214, 265–266.

Dorling, D. (1992). Stretching space and splicing time: from cartographic animation to interactive visualization. *Cartography and Geographic Information Systems, 19*(4), 215–227, 267–270.

Ehlschlaeger, C. R., Shortridge, A. M., & Goodchild, M. F. (1997). Visualizing spatial data uncertainty using animation. *Computers and Geosciences, 23*(4), 387–395.

Fisher, P. (1994). Animation and sound for the visualization of uncertain spatial information. In H. M. Hearnshaw & D. J. Unwin (Eds.), *Visualization in geographical information systems* (pp. 181–185). Chichester, UK: Wiley.

Goldsberry, K., & Battersby, S. (2009). Issues of change detection in animated choropleth maps. *Cartographica, 44*(3), 201–215. doi:10.3138/carto.44.3.201

Griffin, A. L., MacEachren, A. M., Hardisty, F., Steiner, E., & Li, B. (2006). A comparison of animated maps with static small-multiple maps for visually identifying space-time clusters. *Annals of the Association of American Geographers, 96*(4), 740–753. doi:10.1111/j.1467-8306.2006.00514.x

Harrower, M. (2004). A look at the history and future of animated maps. *Cartographica, 39*(3), 33–42.

Harrower, M. (2007a). The cognitive limits of animated maps. *Cartographica, 42*(4), 349–357.

Harrower, M. (2007b). Unclassed animated choropleth maps. *The Cartographic Journal, 44*(4), 313–320.

Harrower, M., & Fabrikant, S. (2008). The role of map animation in geographic visualization. In M. Dodge, M. McDerby, & M. Turner (Eds.), *Geographic visualization: concepts, tools and applications* (pp. 49–65). Chichester, UK: Wiley.

Jiang, B. (1996). Cartographic visualization: analytical and communication tools. *Cartography, 25*(2), 1–11.

Kraak, M.-J. (2007). Cartography and the use of animation. In W. Cartwright, M. P. Peterson, & G. Gartner (Eds.), *Multimedia cartography* (pp. 317–326). New York: Springer.

MacEachren, A. M. (1994). Time as a cartographic variable. In H. M. Hearnshaw & D. J. Unwin (Eds.), *Visualization in geographical information systems* (pp. 115–130). Chichester, UK: Wiley.

MacEachren, A. M. (1995). *How maps work: representation, visualization, and design* (p. xiii, 513 p.). New York: Guilford Press.

MacEachren, A. M., & DiBiase, D. (1991). Animated maps of aggregate data: conceptual and practical problems. *Cartography and Geographic Information Systems, 18*(4), 221–229.

Monmonier, M. S. (1990). Strategies for the visualization of geographic time-series data. *Cartographica, 27*(1), 30–45.

Peuquet, D. J. (1994). It's about time: a conceptual framework for the representation of temporal dynamics in geographic information systems. *Annals of the Association of American Geographers, 84*(3), 441–461.

Peuquet, D. J. (2002). *Representations of space and time*. New York: Guilford Press.

Slocum, T. A., & Egbert, S. L. (1993). Knowledge acquisition from choropleth maps. *Cartography and Geographic Information Systems, 20*(2), 83–95.

Thrower, N. J. W. (1959). Animated cartography. *The Professional Geographer, 11*(6), 9–12.

Tobler, W. (1970). A computer movie simulating urban growth in the Detroit region. *Economic Geography, 46,* 234–240.

Ware, C. (2008). Visual thinking for design. In S. Card, J. Grudin, & J. Nielsen (Eds.), *The Morgan Kaufmann Series in Interactive Technologies* (p. 197). New York: Morgan Kaufmann Publishers.

Weinschenk, S. M. (2011). *100 things every designer needs to know about people.* Berkeley, CA: New Riders.

11

Sound and Touch

Introduction

Think about it: When using your smartphone, tablet, or computer, what normally catches your attention first? Following animations, it is probably one of the following: sounds and vibrations. Sound might include a beep, ring tone, or tune that plays when new information arrives on your phone. Vibration may occur when you select an incorrect option in an app. Sound and vibration are two attention grabbers that can also be used in mobile Web map design. In this chapter, we review a few of these less-talked-about nonvisual map components that can help facilitate Web map communication.

Sound Variables

Web mapping allows for the use of sound to reinforce visual variables or represent data in its own right. When digital cartography began to take off in earnest in the 1990s, sound was considered a potential game changer. It was thought that it would be useful for supplementing visual variables. In the early days of Web cartography (i.e., 1990s and early 2000s), many Web maps made extensive use of sound (e.g., clicking sounds when buttons were pressed, music in the background, etc.). The rise of Macromedia Flash allowed Web cartographers to embed all sorts of noises in their maps with minimal effort. Today, however, fewer Web maps incorporate sound. (The reasons why are reviewed in this chapter.)

Krygier's Sound Variables

Krygier (1994) was one of the earliest researchers of the use of sound as a map variable. He proposed and tested nine sound variables that could potentially be used to signify spatial data. It turns out that, for the most part, many of the variations possible with sound are not very applicable to maps. For example, human hearing is incapable of inferring quantitative differences between sounds, so highlighting specific data differences with sound is difficult. Even with these shortcomings, though, sound can prove beneficial. This is particularly true when used to *enhance* visual variables or to *highlight ordinal differences* among data values. Of potential interest in Web mapping are the following variables: loudness, pitch, duration, and attack/decay.

Loudness

Loudness (i.e., volume) can be used to represent more or less of something. For example, loudness is an excellent sonic variable for depicting different levels of data density. If you are designing a Web map with a WIMP (windows, icons, menus, pointer) interface, you could include loudness in your map so that as someone mouses over different enumeration units, the volume goes up and down depending on the quantity of data found therein. The sound can also be tweaked to better mimic what is being represented. If you are designing a map of population density, for example, the sound used could be that of a group of people talking. The more densely populated a place is, the louder and more intense the talking could be. The less-densely populated, the quieter and softer the talking would be.

Pitch

Less useful, perhaps, and potentially more painful for pets to listen to, is pitch. Pitch allows someone to determine the value of a map object by associating the pitch of a sound within a range of pitches. Similar to how an unclassified choropleth map allows someone to obtain a rough estimate of where an enumeration unit's value lies among others, pitch can do the same sonically. Unlike an unclassified choropleth map, however, one cannot simply look at other enumeration units or data points to compare pitches to one another. Sound variables require interacting with numerous data points or engaging with an interactive legend so that comparisons can be made. Thus, if pitch is used, it is probably best done in conjunction with a visual variable of some sort. For example, if you were mapping crime rates by neighborhood using a choropleth technique, you might add an ascending-pitch variable. When someone mouses over a high-crime neighborhood, a high pitch representing the value of that class might be played.

Duration

Duration represents how long a sound is, *or is not,* perceptible. This has limited use in Web mapping, but some creative uses can be envisioned. For example, using this sonic variable to complement a visual variable (e.g., perspective height) when one moves the mouse over an object may help map users better interpret the data. One could use the sound of something going up (a sound akin to one of those test-your-strength, sledgehammer games at a carnival). The longer the sound lasts, the higher the value will be.

Attack and Decay

Attack and decay offers an interesting sound variable for representing time. Attack and decay refers to the length of time it takes a sound to reach its highest volume and its lowest volume. One potentially effective use of attack and decay is to represent temporal data on a static map. For example, attack and decay could be used to represent the length of different world empires by having sounds play and gradually disappear as empires come and go.

Sound Decay?

Although they have been known and tested for around two decades, sound variables are still not frequently employed to represent spatial data on maps. One reason for this is that computer and Web technologies are changing. In the 1990s, when sound was being rigorously investigated as a method of data communication, personal computers did not yet function as multimedia centers. There were no YouTube, Hulu+, Netflix, iTunes, eBooks, app stores, or other online multimedia sources. Today, most people surf the Web on mobile devices, many while concurrently listening to music in the cloud or watching online video of some sort. Even when not on a mobile device, computers have become entertainment hubs; they act as televisions, radios, music collections, and video game consoles. In essence, the soundscape of society has changed, and it is increasingly likely that someone will be using a Web map and listening to something else on his or her computer while doing so. Having a map that attempts to differentiate spatial data via different levels of noise while you are busy jamming to a Radiohead album is therefore not all that useful.

Nevertheless, sound can be effective when designing map interfaces. Sounds can reinforce interaction on digital devices by mimicking real-world sounds. For example, when someone clicks on a touchscreen keyboard, the use of sound can inform the person that the letter has been pressed. Likewise, when someone selects an icon on a map, a sound can be used so that the map user understands he or she has successfully selected an object.

This is not to say that sound should never be employed as a map data variable. In fact, it is my belief that one thing that is missing from contemporary Web maps is the effective use of sound hints and ambient music. (It adds a lot to video games—why not maps?) However, you need to take into account your primary and intended audience. If you think the majority of people are going to be using your map while sitting on the couch streaming music from their device through their Bluetooth stereo, then it is better not to include sound. On the other hand, if your map is going to be used by specialists to analyze the demise of blue heron habitat in northern Minnesota, then sound may add a lot.

Tactile Variables

Tactile maps—maps that use aspects of touch to convey information about spatial data—have been studied for decades. Beginning with the exploration of making maps for the blind, tactile variables have been used to create everything from reference maps, to show elevation, and even represent thematic data. Digital technology and mobile smart devices, however, are making tactility even more useful for mapmakers as they add the benefit of user-device interaction. (They are also coming a long way toward making Web maps more accessible for the blind. For an example, see the YouTube video referenced at the end of this chapter in "Further Reading and Resources.")

Numerous tactile map variables have been proposed throughout the years. However, most of these tactile variables are not relevant to Web cartography as they are dependent on mutation of a map's medium. Digital screens are not as malleable as plastic or paper. Thus, tactility in relation to Web mapping rarely deals with anything more than the vibration of the device being interacted with. Such map interactivity is increasingly referred to as haptic.

Haptic variables can include everything from vibrations to a pinching sensation. Although the use of vibration and its potential effectiveness for Web mapping are still largely unexplored, map users are already used to haptic interaction with their devices by default. When we type something on a touch device keyboard, we are often greeted with a small vibration indicating we have successfully typed a letter. This also can be useful on a map when selecting items in an interactive smart legend or attempting to highlight a particular dataset the map user has selected.

There is an important caveat to including vibration as a major component of your Web map. If you are designing a Web map for both WIMP and post-WIMP devices, vibration may not be perceptible for many of your map users. Most computer monitors, personal computer (PC) devices, and high-definition televisions (HDTVs) do not have vibration functionality, although this may change in the future. Finally, vibration should only be used to supplement important visual cues. If it is overused, it can detract from the mapped information. Remember, design with a purpose. If it does not really add anything to the design or intuitiveness of the map, do not use it.

Key Concepts

- Sound can be successfully included in maps to add value to the data being mapped.
- Never forget that sound on maps will likely be competing with other types of multimedia concurrently being run on a map user's computer or mobile device. Thus, the sounds may not be heard, or may have been intentionally muted, when someone is using your map.
- The usefulness of sound for presenting information to map users is largely limited. However, loudness, duration, pitch, and attack and decay may prove useful for supplementing visual stimuli.
- Haptic feedback (such as mobile device vibration) is still largely limited to touchscreen devices. However, it does offer some opportunities for providing map users additional information.

Further Reading and Resource

Resource

An example of haptic and sonic map feedback used to create Web maps accessible for the blind is available at http://goo.gl/R1a4B

Further Reading

Brauen, G. (2006). Designing interactive sound maps using scalable vector graphics. *Cartographica: The International Journal for Geographic Information and Geovisualization, 41*(1), 59–72. doi:10.3138/5512-628G-2H57-H675

Caquard, S., Brauen, G., Wright, B., & Jasen, P. (2008). Designing sound in cybercartography: from structured cinematic narratives to unpredictable sound/image interactions. *International Journal of Geographical Information Science, 22*(11/12), 1219–1245. doi:10.1080/13658810801909649

Fisher, P. (1994). Animation and sound for the visualization of uncertain spatial information. In H. M. Hearnshaw & D. J. Unwin (Eds.), *Visualization in geographical information systems* (pp. 181–185). Chichester, UK: Wiley.

Krygier, J. B. (1994). Sound and geographic visualization. In A. M. MacEachren & D. R. F. Taylor (Eds.), *Visualization in modern cartography* (pp. 149–166). Oxford, UK: Pergamon.

12

Web Map Production

Introduction

Technology changes rapidly and not always in ways we can predict. Many years ago, while in graduate school at Penn State, I began exploring this new-fangled thing called JavaScript. Everything I was hearing about it was largely negative. Conventional wisdom was that learning JavaScript was likely a waste of time. It did not work in all Web browsers. Even in the browsers where it could work, many people intentionally disabled it. Programmers ridiculed JavaScript for not being a "real" language—as JavaScript is not compiled. Macromedia Flash (which is now Adobe Flash) was heralded as the future of online interactivity and mapping. I subsequently sold my JavaScript guidebook and bought a Macromedia Flash one. Oops.

As the world increasingly accesses Web and multimedia content from screens the size of playing cards, Adobe Flash Player has been relegated to PC browsers. JavaScript is the king of Web mapping. Today, the conventional wisdom is that you must learn JavaScript to make Web maps. This is both good and bad news for mapmakers. It means that we probably have another 5–10 years of using this technology to create stunning Web maps. Alternatively, JavaScript's ascendency means that it is already starting to use up its 15 minutes of fame. No programming language stays in the limelight forever.

Interactive design technologies like Adobe Flash and JavaScript are only but a piece of the Web mapping puzzle, however. New Web map services (e.g., Google Maps, OpenStreetMap) have arrived on the scene and continue to evolve rapidly. New spatial data types have arisen that more readily lend themselves to being mapped on the Web (e.g., KML [Keyhole Markup Language], GeoJSON [so named because it represents geographic data in JavaScript Object Notation], map tiles). Additional vector and raster technologies abound that allow mapmakers to take a static map designed for print and quickly turn it into an interactive map for distribution on the Web. Web maps are no longer confined to a browser. It is now possible to package interactive maps as mobile apps, or *mapps*, to be sold just like paper maps.

Many of these technological innovations are wonderful; they have resulted in a cornucopia of new maps and cartographic experimentation. Yet, at the same time, for those just starting with Web mapping, all of these new data types, technologies, and tools can be overwhelming. What if you just want to design a simple, effective Web map? Where do you start? What technologies are most suitable for your needs? On the other hand, if you are overseeing the development of a large, longitudinal mapping project, how do you make sure you are using the technologies that will likely have the most resilience and reliability for years to come?

The rest of this chapter is an attempt to explain in layperson terms (1) what you need to know about web coding; (2) new spatial data types for the Web and how to create them, and (3) tools and services for creating Web maps. I have done everything I can in this book not to promote one technology over another. It is my hope that you can use the information in this chapter to explore your different mapmaking and data options and decide which technology is most suitable and practical for your needs.

The Gist on Coding

If you want to design Web maps for a living, or even as a hobby, you have to learn how to read and write some code. The good news is this: You *do not* have to become a coding master. To be sure, the better your coding skills, the easier Web mapping will be. But, with just a little bit of coding skill you can do incredible things. You can create interactive legends and map elements, design intuitive map layouts, and present information in extremely dynamic ways. However, those who have no coding experience should have no fear. All you need to be able to do is learn enough code to hack other people's code into something you can use. Learning how to code actually does not take much time. So, where do you start?

What Was Flash?

Perhaps you have heard the chants, or at least seen the tweets: "Flash is dead!" Though still quite ubiquitous in personal computer (PC) browsers, the Adobe Flash Player plug-in is now mostly absent from mobile devices. It would seem that this technology is truly dying off. In fact, most plug-ins seem to be in a slow death spiral.

A Web plug-in is an external program that is loaded by a Web browser. There are many browser plug-ins, including Adobe Flash Player, Oracle Java, Apple QuickTime, Google Earth, and Adobe Acrobat, to name a few. These programs are used to run content compiled in a language other than HTML

(Hypertext Markup Language, described in more detail later in this chapter). Essentially, they are small programs used to execute external, non-HTML code (i.e., to play videos in your browser). Adobe Flash Player runs SWF (ShockWave Flash) and FLV (Flash Video) files—files typically created using the Adobe Flash Professional or Adobe Flash Builder programs. Adobe Flash Player essentially allowed people to create dynamic, interactive, and animated features for the Internet. This was revolutionary because older versions of HTML were extremely static in nature.

What Is HTML5?

HTML5 is an attempt to create a dynamic Web devoid of plug-ins. The goal is to create a framework of open-source technologies that can be used together to mimic everything that thus far only plug-ins like Flash Player could really do. The belief is that native HTML5 coding and technology will be (1) faster, as external programs do not need to be loaded by the browser; (2) more secure, as exploits that were not easily detected by browsers or antivirus software because they worked through plug-ins will now be largely nonexistent; and (3) more open, as people can feasibly design HTML5 with nothing but a text editor.

HTML5 is a bit of a misnomer. Yes, it is the fifth iteration of HTML. However, the technology is actually comprised of three separate languages: HTML, Cascading Style Sheets (CSS), and JavaScript. Each of these three components, using its own unique coding syntax, has a distinctive role in making Web sites work. You must become familiar with all three of these languages to design Web maps. All additional Web coding languages you might learn will be supplementary to these.

Again, you do not need to become an expert HTML, CSS, and JavaScript coder to make great Web maps. You just need to know enough to interpret other people's code, tweak code that you have copied, and understand how to interact with application programming interfaces (APIs) (discussed also in this chapter). Think of it this way: You do not need to be fluent in a language to visit another country. However, if you want to get around on your own, order lunch, and intermingle with the locals, it is best to take the time to pick up a few key phrases.

HTML

Hypertext markup language is what lends HTML5 its name. HTML has always been the backbone of the Web, the language in which content is provided to Web browsers. The syntax of HTML is quite simple; it works via the use of tags (see Figure 12.1). HTML was traditionally limited when it came to interactivity and design capabilities. HTML5 (which is nearing completion but is not implemented across all Web browsers in the same manner)

```
<head>
<title>This is What HTML Really Looks Like</title>
<script>
    // This is where you could put JavaScript. It goes between two
<script> tags.
    // CSS and JavaScript typically go in the <head> of an HTML
    // page. This is where
    // all the stuff that browser users don't see gets done.
    // JavaScript can be put anywhere in the HTML document, though.
</script>

</head>

<body>

<p> The body represents the heart of an HTML page. It is here that
you organize your content. It will never look pretty written up in
simple HTML. However, you can import or write CSS in the HTML to
stylize and format the content.</p>

<div id="organizational tool"> Divs are tags that allow you to
clump a variety of things together. They can then be styled or
arranged on the page using CSS. These are great for menu and map
element creation, as you will have control over their style and
placement with CSS. It will also allow you to interact with these
using
JavaScript.</div>

<p>That's it for now!</p>

</body>
</html>
```

FIGURE 12.1

A sample of what HTML code looks like. Notice the tag format (<tag> ... </tag>). Everything is nested within tags. A browser simply goes through and reads these tags to interpret the content.

has been revised to include a whole slew of new interactive and dynamic features that were once the domain of external browser plug-ins, such as the Adobe Flash plug-in.

HTML is now used almost solely for content organizing and delivery. In the new HTML5 system, HTML is used for content. Styling and interactivity are left to other languages—CSS and JavaScript.

Cascading Style Sheets

Whereas HTML is used to provide and *organize* content sent to a Web browser, CSS *styles* this content within the browser. CSS is nothing more than a styling language. Its whole purpose is to format markup languages, namely, HTML and XML (Extensible Markup Language) (discussed later in the chapter). It has existed since 1996, yet its implementation by different Web browsers has been slow and haphazard. Different Web browsers are able to interpret different amounts of CSS code, and they frequently interpret it differently. This is finally starting to change with the most recent iterations of Web browsers, but browser differences continue to make Web design far less predictable than it could be.

CSS is what you will need to learn to make your Web maps attractive. It is the language used to design the look and feel of Web map elements: borders, fills, outlines, drop shadows, shapes—everything. You will even use CSS to design your map layouts. Anywhere there is HTML or XML content, you can use CSS to style it to your specifications.

Fortunately, CSS is fairly easy to write and comprehend. It is written more like a list of characteristics than a language. Essentially, you write a list of rules. The browser interprets these style rules and reformats the HTML accordingly. All of the rules are written in the same two-part syntax (see Figure 12.2). You write a "selector" and then follow it with a "declaration." A selector, as its name implies, selects which part of the HTML code is going to be styled. The declaration simply declares what properties of the selected HTML elements will be styled differently from the default HTML style and assigns them new values. Obviously, CSS can become complex and is not always as straightforward as the examples presented in Figure 12.2. Due to its simple structure, however, CSS is likely the easiest language of the three comprising HTML5 to comprehend and write.

JavaScript

HTML is used for content organization and CSS is used for content styling. So where does the interactivity come from? Enter, JavaScript. JavaScript is an object-oriented scripting language that is used to target specific, predefined objects (e.g., map elements) and manipulate them under certain defined conditions.

Unlike plug-in-based scripts, JavaScript is read and run by the Web browser itself. It is known as a client-side script because the processing is done on a user's computer. The benefit of this is that map data processing on a map user's machine is typically faster than processing on a server before sending it to the computer. Moreover, JavaScript is far more responsive to user interactivity with your map because the browser itself is processing user input.

JavaScript is used to write functions, or methods, that directly manipulate HTML content. (HTML content that can be interacted with is frequently referred to as the Document Object Model, or DOM, so when you inevitably see this acronym, do not panic.) JavaScript can do all sorts of things, such as allow you to interact and update Web page content without reloading the page, animate parts of a Web map, validate form information, and fetch external data. Essentially, if you are using HTML5 for your Web maps, all interactivity in your Web map will be created with JavaScript.

JavaScript is a very different language from HTML or CSS. HTML is a tag-based language. Everything is nested in tags. CSS is written as a series of style

```
body {
    background-color: #000000;
    margin: 0;
    padding: 0;
    color: #000;
    font-family: Corbel, "Myriad Pro",
Calibri, Verdana, Arial, Helvetica,
sans-serif;
    font-size: 100%;
    line-height: 1.4;
    cursor: crosshair;
    page-break-before: auto;
    page-break-after: auto;
}
.phoneImage {
    right: auto;
    position: static;
    top: auto;
    -webkit-transition: all 5s;
    -moz-transition: all 5s;
    -ms-transition: all 5s;
    -o-transition: all 5s;
    transition: all 5s;
}
ul, ol, dl {
    padding: 0;
    margin: 0;
}
h1, h2, h3, h4, h5, h6, p {
    margin-top: 0;
    padding-right: 15px;
    padding-left: 15px;
}
```

FIGURE 12.2

CSS is a litany of code that typically looks similar to this. CSS rules always have a selector (typically an HTML element, such as "body"). The selector is always followed by a declaration in curly braces. For example, this CSS selects the Web page's body tag and declares that the background color property will be black, and that the fonts the Web page should use, in order of preference, are Corbel, Myriad Pro, Calibri, Verdana, Arial, Helvetica, or a default sans-serif text.

rules, each with a selector and a descriptor. JavaScript is an object-oriented programming (OOP) language. It is quite robust. With JavaScript, you can create new variables (or objects) and functions (or methods that do things to objects), and you can create loops and conditionals so that functions are only run on certain objects or when certain parameters are met. Better yet, you can write a script that does something—say, creates a button that minimizes all open map elements except the mapped area when it is clicked—and share it with others or reuse it on all of your maps (see Figure 12.3). You can write code to add a particular bit of interactivity to a map once, and it will be reusable and modifiable forever.

Although JavaScript is far too complicated to discuss in detail here, what needs to be emphasized is that the majority of online Web mapping

```
<script type="text/javascript"
src="https://maps.googleapis.com/maps/api/js?key=APIKEYensor=true">
    </script>
    <script type="text/javascript">
      function initialize() {
        var mapOptions = {
          center: new google.maps.LatLng(-34.397, 150.644),
          disableDoubleClickZoom: false,
          zoom: 8,
          panControl: false,
          overviewMapControl: true,
          mapTypeId: google.maps.MapTypeId.ROADMAP,
          streetViewControl: false,
          mapTypeControl: google.maps.MapTypeControlStyle.DROPDOWN_MENU,
          zoomControlOptions: {
          style: google.maps.ZoomControlStyle.SMALL,
          position: google.maps.ControlPosition.LEFT_TOP,
        },
        infoWindowOptions: {
            pixelOffset: 3,
            maxWidth: 15,
        }
    };
        var map = new google.maps.Map(document.getElementById("map_canvas"),
            mapOptions);
        var farmersMarkets = new
google.maps.KmlLayer('http://www.ian.muehlenhaus.com/KML/farmersMarket.kmz');
        function loadKML(){
            farmersMarkets.setMap(map);
        }
        loadKML();
    }
    </script>
```

FIGURE 12.3

This is an example of JavaScript code that you will see if you use any map API. JavaScript can be written in a separate text file and imported into an HTML document, or it can be written within the HTML document itself between two script tags. The first <script> tags on this page import the Google Maps JavaScript API. The second set of <script> tags actually run several functions to initialize the map on the Web page, tweak the interface a bit (no panning arrows), and add a KML layer showing farmers' markets. Although learning JavaScript may appear daunting at first, all of this code was simply copied, pasted, and tweaked to suit this figure's needs straight from the Google Maps API tutorials site. You will not need to be an expert at JavaScript; you just need to learn enough to borrow and hack free code that is already out there.

is currently being done using JavaScript to one extent or another. I highly encourage anyone serious about Web mapping to start learning how to read, write, and hack JavaScript code. The JavaScript mapping community is huge. Online, you can find thousands of open-source mapping "scripts" that are free for the taking. You can import these into your HTML and start doing marvelous things with your Web maps.

Application Programming Interfaces

I know that I said HTML5 is comprised of three parts. So, what is this fourth, API thing? Modern Web mapping makes heavy use of APIs. APIs are basically communication protocols that let different computer programs speak and interact with one another. In the case of Web mapping, APIs typically allow the browser to use JavaScript to communicate with an online mapping service. If you needed one more reason to start learning JavaScript, as if Web interactivity and animation were not enough already, it is so you can start playing with third-party APIs in your Web maps.

For example, Microsoft Bing Maps is an incredible resource. They have all of this great mapping data. They could just collect all of this and bottle it up on their Bing Maps site, requiring anyone who wants to use their mapping technology to visit a stand-alone Web site. Instead, though, they facilitate wider use of their map service by allowing Web programmers to interact with their spatial databases to make their own specific Web map applications. To do this, Microsoft Bing Maps has to set up an API. This is a predetermined set of coded instructions (i.e., the protocol) designed by Microsoft. Designers can then call on this protocol in their Web sites and interact with it (again, typically via JavaScript). Using protocol codes and guidelines set up by Microsoft, map designers can access, import, manipulate, and design Bing Maps and map data within their Web site.

I used Microsoft Bing Maps as an example here because I really like their base map design. However, there are hundreds of APIs out there. Nokia, Bing, Google, MapQuest, Leaflet, and many other APIs are designed for interacting with different map services. Not all have to do with mapping, however. There are APIs for all sorts of online services, including photo Web sites (e.g., Flickr), social networks (e.g., Facebook), and productivity services (e.g., Google Drive). What is truly awesome is that a Web site can incorporate and use many APIs at the same time, combining the power of many of these services in one Web site or mobile app.

Prepping Spatial Data for the Web

Regardless of the Web technology you choose to produce your maps, you will need to figure out how to create and export your spatial data in a format that is suitable for Web mapping. In this section, I review characteristics of several different spatial data options available. First, I briefly discuss some of the most common data formats used in Web mapping. I then conclude with some software recommendations for exporting your GIS and non-Web format spatial data into Web-ready formats.

Choosing a Projection

If you decide to use another service's API, be assured that you will probably want to prepare your data for the Web Mercator projection. As mentioned in this book previously, for better or worse, the Web Mercator projection has become the standard Internet projection. Although many APIs allow you to change the projection via coding, it is safe to assume that generally you will be coerced into the Web Mercator (see Figure 12.4).

As discussed in this book, however, not every map should be designed using the Web Mercator projection. I highly recommend not using an API if it is forcing you to use this projection with which your data cannot be adequately displayed. This is particularly true when creating small-scale thematic maps. As will be discussed further here, designing a Web map using appropriately projected Scalable Vector Graphics (SVG) is preferable to an API in these cases.

FIGURE 12.4

The specter of Mercator haunts Web mapping. Again, Greenland is roughly the size of Mexico—not so on this map. In fact, on this map Greenland has the same area as Africa; in reality, Africa is 14 times larger. Areas become grossly distorted the farther they are from the equator. (©OpenStreetMap contributors; available for free under the Open Database License.)

Common Web Map Data Formats

Depending on the software and scripting languages you are using to make your Web maps, there are innumerable spatial data formats at your disposal. Many of these Web data formats are not what print cartographers and desktop GIS (geographic information system) users are used to seeing. Next, I outline some of the most prominent data formats used by Web cartographers.

Keyhole Markup Language

Keyhole Markup Language is an XML language developed by Google to represent spatial data. XML is simply another tag-based language similar to HTML. What makes XML unique is that people can label and name their tags in any way they wish. This is one reason it has become so wildly popular. It is a language with syntax but no vocabulary until you invent it yourself. KML files are written using particular XML coding that can be read by Google Earth and many other mapping services. KML files are easy to create and, since they are merely XML files with special notation, can be written and edited using a simple text editor. Although many APIs do not have the means to import KML as map overlays directly, numerous third-party JavaScript codes have been written that allow for the easy importation of KML into almost any mapping project.

You will also come across KMZ files. These are zipped KML files. KMZ files contain a base KML file as well as supplemental materials to which the KML file refers (e.g., photos, icons, overlays).

GeoJSON

GeoJSON is becoming one of the most common spatial data types used in Web mapping. Its popularity arises from several factors: It results in slightly smaller file sizes as compared to KML and is considered by some to be more intuitive than XML for reading and editing. This spatial data type is extremely easy to control and manipulate using JavaScript, which has resulted in its easy importation in many APIs.

Scalable Vector Graphics

An SVG is an XML image format (i.e., the graphic is saved as an XML text file). SVG items do not pixelate with resizing, unlike traditional raster image files, because they are created using coordinates in an XML file (see Figure 12.5). Thus, regardless of how far someone zooms in or out on an image, the points and lines comprising the graphic are redrawn at an appropriate resolution.

SVG items can be created using many different tools, including Adobe Illustrator and several online resources. An excellent feature about SVG files is that you can manipulate and animate them using JavaScript. SVG files are stellar for creating non-Web Mercator thematic maps.

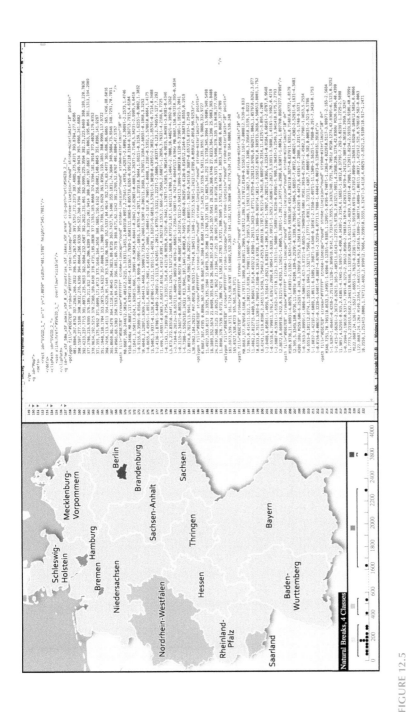

FIGURE 12.5

The map on the left is an SVG of the natural breaks figure found in Chapter 9. The SVG was exported from Adobe Illustrator. The picture on the left was taken when the SVG was opened in the Mozilla Firefox Web browser. The image on the right is the exact same SVG file opened in a text editor instead of a browser. SVG files are actually XML text files. Certain programs can read and translate them into graphics, including most Web browsers.

Map Tiles

Map tiles are an effective way to distribute detailed base maps that are generalized for numerous scales on the Web. Map tiles work by breaking a large map down into rasterized squares—or "tiles"—that can be loaded individually by a Web browser. These tiles are commonly sized at 256 by 256 pixels. The benefit of breaking down a large base map into tiles is that if someone zooms in on the map, the browser can spend its resources loading only those tiles that need to be shown on the map user's screen. It can ignore loading all of the other tiles in the map. The browser can also cache adjacent, off-screen tiles should a user want to pan the map. This results in very fast, seamless map loading times.

Importantly, as someone zooms in on a map, different, more detailed tile sets can be loaded. For example, at a scale of 1:80,000,000, a map user might only be shown country boundaries and a few country names on the tiles. As the map user zooms in—making the map a larger scale, for example, 1:100,000—cities, roads, and highways can start to be shown on a more detailed tile set. On many slippy maps, it is common for 15–20 different sets of tiles for the same area to be made, with different scales and levels of generalization. Each zoom level loads a different map, broken down into hundreds or thousands of 256×256 pixel tiles.

Another use of map tiling is to break down a high-resolution raster image of a classic map so that a browser can load it more rapidly. This can be used to make viewing a very large, high-resolution image faster and easier in a browser. By breaking a 2-gigabyte image down into tiles, for example, when a user zooms in on the map to see more details, only those tiles filling the browser screen need to be loaded—not all 2 gigabytes.

Tile maps are incredibly effective, but they do suffer from several drawbacks. First, to host detailed tiles of large areas (e.g., a continent or the world) takes a massive amount of server space. Storing so much data online for a hemispheric or global map is costly. For an interesting analysis of how much tile data Google must serve over the Internet to map the world in the amount of detail it does, I refer you to Michael Peterson's (2011) article "Travels with iPad Maps" (see "Further Reading" at the end of this chapter), which is freely available online. The expense of serving up one's own tile data is one reason it is often a great idea to use another person's hosted tiles (e.g., OpenStreetMap, further discussed later in this chapter). Many APIs now make it easy to style the appearance of their tiles. This is often a great option instead of creating and hosting your own set.

A second drawback of using your own tiles is that if something drastic changes with your base map (e.g., Atlantis finally shows up in the middle of the ocean), it can be time consuming to update the map. It requires going into your map creation software, changing the base map, reexporting the map into different tiles, and then uploading them again.

Geospatial PDFs

Geospatial PDFs, or GeoPDFs, are georeferenced maps in a portable document format (PDF). These maps have been commonly used and distributed by the U.S. military, as they open in Adobe Acrobat or any PDF viewer as well as a host of mapping applications. GeoPDFs can include additional GIS data such as layers and icons. Because they are georeferenced and can work in conjunction with a device's GPS, they are useful on mobile devices. They are typically downloaded to and stored on a device; therefore, no Internet connection is needed to interact with and find distances on these maps. This is ideal when needing a map in a remote location without a reliable Internet connection. Finally, since these are typically paper maps turned into pan-and-zoom Web maps, the look and style of these maps is often pleasing for those of us who grew up with static maps.

The downside of GeoPDFs is that they are not as widely adopted as many other file and data types. Avenza has created mobile apps that allow one to download and interact with GeoPDFs (see Figure 12.6). Avenza PDF Maps even has an integrated market where you can find, sell, and buy GeoPDF maps for immediate consumption.

How Do I Create These Data Types?

How do you create GeoJSON files? Where can you go to export Web map tiles? How do you find or create data-rich KML files? GeoPDFs sound cool; why is there no option to export to this format from my GIS?

I have good news: There are innumerable ways to create data in these formats. Most of the software to do so is free. If you have some of your budget to spend on software, though, you are truly in luck. There are some really versatile mapping programs that are worth the money simply for the convenience they afford you in creating, generalizing, and exporting spatial data to suit your specific needs.

Geographic Information Systems

Many GIS software packages allow you to export data into different data formats that are suitable for Web mapping. Unfortunately, which formats they allow you to export your data into vary greatly. Due to the proprietary nature of many GISs, they do not always allow you to export into all of the file formats you desire. Some common GIS software packages are Blue Marble Geographics Global Mapper (proprietary), ESRI ArcInfo (proprietary), and Quantum GIS (open and free).

Blue Marble Geographics Global Mapper

Personally, one of my favorite programs is Blue Marble Geographics Global Mapper software. This GIS and data management application imports and

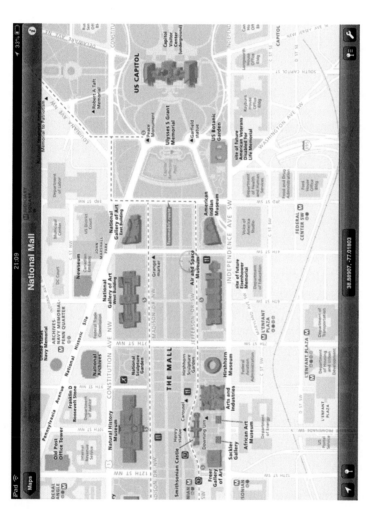

FIGURE 12.6

This is a screen capture from iPad2 of the National Park Service (NPS) National Mall Map. This map is a GeoPDF. It retains all of the characteristics of the NPS's high-quality paper maps, but it is GPS enabled and has pan and zoom features. The map interface is Avenza's PDF Maps for iOS app. Using this app you can download many free maps to store on your mobile device. Importantly for map developers, you can distribute your maps to iOS devices through Avenza's PDF Maps market. An Android version of PDF Maps was in beta testing at time of publication. (App Copyright: Avenza. Used with permission. Map: National Park Service.)

exports an "unparalleled variety of spatial datasets" (http://www.bluemar-blegeo.com). You can import almost any type of spatial data file, regardless of whether it is open source or proprietary, designed for desktop GIS applications or Web maps. It also has built-in connections to map servers, allowing for quick, easy raster data download of elevation datasets, topographic map collections, and more. You can then manipulate and export these data into almost every single geospatial data format that exists, including every style mentioned in this chapter and over 100 more. For example, if you want to take an out-of-copyright map, georeference it, and export it as a series of tiles for a Web map, Global Mapper can do it in a snap. Or, if you want to take an ESRI Personal Geodatabase file and export it as a GeoJSON, it can do that too. It is an exceptional program for managing data, in addition to being useful for a variety of other analytical and standard GIS features. It is also afford-able by GIS standards.

ESRI ArcGIS Desktop

Most people who come to Web mapping from a GIS background have extensive experience with ESRI ArcGIS Desktop (http://www.esri.com). This program is also capable of organizing, projecting, and exporting your data for the Web. In fact, the latest iterations of the software are starting to muddy the waters between desktop and Web mapping, as the desktop software is very much tied into online databases and Web services. If you plan to use the robust and relatively user-friendly ArcGIS API (discussed later in this chapter), then ESRI ArcGIS Desktop will likely be your data tool of choice.

Although Blue Marble Geographics Global Mapper and other GIS software packages such as ESRI ArcInfo can speed up data processing and organization, they do have a downside: They cost money. Fortunately, free services and open-source programs that allow you to convert data from one file format to another abound on the Internet.

Quantum GIS

Quantum GIS is free GIS software that runs on Mac, PC, UNIX, Linux, and Android systems (http://www.qgis.org). It is user friendly and surprisingly powerful. There is a large community of users, and the software has really come a long way in recent years in replicating the power of proprietary GIS software. It is definitely worth installing if you are looking for a free spatial data manipulation and analysis tool.

Non-GIS Data Creation Applications

You absolutely do not have to use GIS to make Web maps. You can write HTML, CSS, and JavaScript code using a default text editor and upload all of your files to a Web site using an operating system's built-in FTP tool. You can also type your KML and GeoJSON files in a text editor.

However, there are even simpler solutions than this. If you have access to the Internet, you can start using free cloud-based Web mapping services that not only create basic-looking maps for you but also allow you to export your data into a slew of different Web formats. I review some of these sites here, although new ones are being developed all of the time.

CartoDB

CartoDB (http://www.cartodb.com) is a wonderful, free cloud mapping service. Of pertinence, though, is the fact that you can upload all sorts of spatial data, style it, and then export it into a variety of Web-friendly formats for use in another API (see Figure 12.7).

Geocommons

Geocommons (http://www.geocommons.com) is a great Web site on which to create basic thematic Web maps. More important, however, it can be used to find and convert data. On this Web site, users upload datasets and then share them with the public. You can download these data in a variety of

FIGURE 12.7

CartoDB is an extremely powerful and intuitive mapping service. In addition, it works as a great, free online tool for data conversion. Simply upload one of many different spatial data types and then choose to export your data as a different kind of spatial data type. (Copyright CartoDB. Used with permission.)

formats, including KML. This is also a great place to upload a shapefile and download the KML version to embed in an API.

Google Maps

You probably already know that Google Maps (https://www.maps.google.com) is a great tool for getting directions from your hotel to the nearest coffee house, but it is also useful for Web cartographers making data from scratch. Using the "My Places" feature, you can digitize your own data over Google Map's base map and aerial photos, link attribute and multimedia to your data, save the dataset, and then export it all as a KML file (see Figure 12.8). These KML files in turn can be uploaded to the Web and loaded via API into your own, Google or non-Google, Web maps.

Indiemapper

If you are looking for a service that will accept shapefiles, quickly reproject them, and export them as SVGs, Axis Maps Indiemapper (http://www.indiemapper.com) is your site. Once upon a time, it was a subscription-based mapping platform for designing print maps. Now, it is free. One benefit for Web mappers is that you can upload all of your map data to this site, reproject them, and then export all of the data as a layered SVG file. Then, you can simply use HTML, CSS, and JavaScript to manipulate the SVG map and make it interactive.

MapTiler

MapTiler (http://www.klokantech.com/maptiler/) is a program that takes your data and turns it into map tiles that can then be presented using an API or as a stand-alone slippy map on your Web site. It has a very intuitive graphical user interface (GUI; see Figure 12.9). The free version embeds a watermark in your tiles. For a nominal fee, you can unlock more advanced features and eliminate the watermark.

OpenStreetMap Data and Tiles

OpenStreetMap (http://www.openstreetmap.org) is the Wikipedia of mapmaking. Its spatial data is contributed by individuals around the world using their own intimate local knowledge and GPS data collections. People upload data all the time, and the database is ever expanding. Better still, anyone can use this spatial data for free. For many countries around the world (e.g., Haiti), these data are the most detailed and reliable available, perhaps even better than any official government maps.

OpenStreetMap data can be exported for use in a variety of ways. First, it can be embedded directly into your Web site using an HTML iFrame tag. You can also export the data in OpenStreetMap XML format, which can be opened and exported into a different file format, such as SVG, using several programs (e.g., Global Mapper). Finally, you can simply export the data as an

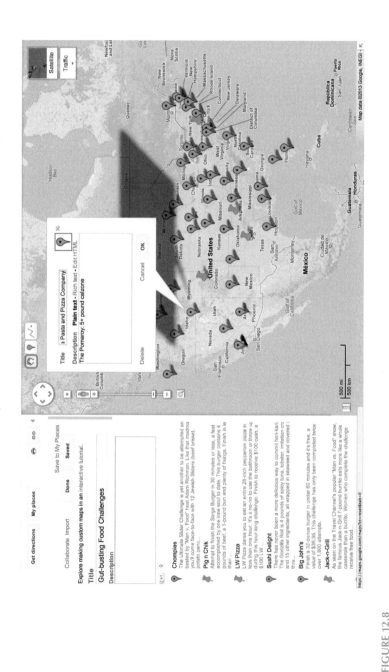

The Google Maps My Places tool is excellent for creating lots of data in the cloud. Here you can digitize (i.e., trace with spatial attributes) points, lines, roads, and polygons. You can manipulate and select color schemes and symbols, add information and data to the info windows, and even embed multimedia (e.g., YouTube videos). When you are done, you hit save, and Google keeps your data as a KML file. Log out and log back into Google Maps to have the option to download the KML file. Once you download it, it is yours to upload and use however you like, with all of the info window data saved. This is a map that one of my Maps 101 students made of "Gut-Busting Food Challenges" around the United States. (Copyright 2013 Google.)

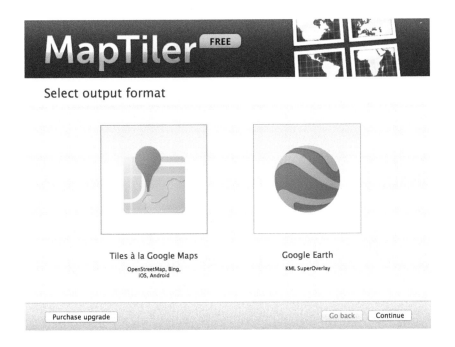

FIGURE 12.9
Map Tiler is a great and intuitive program for breaking your maps down into numerous tile layers. The free version has a watermark, but for a nominal fee you can buy the full license and make as many tile maps as you could ever want. (Copyright 2013 Klokan Technologies. Used with permission.)

image file. Probably most useful is the fact that a handful of other APIs allow you to import tiles directly from OpenStreetMap.

R

The R Project (http://www.r-project.org), also more commonly referred to as R, is a free, open-source programming environment. It was designed to facilitate statistical analysis. However, it does much more than compute numbers. A large user base and many dedicated supporters of the software have created hundreds of packages that can be imported into the program. These packages can do myriad things, everything from running advanced statistical analysis to producing Chernoff faces. One can use R to export map graphics, create maps right in R, and even implement Color Brewer color schemes.

The downside is that R is largely command line in nature. The coding is simple to figure out, but if you are someone who likes programs with GUIs, the learning curve may be high. However, with a few simple tutorials (visit http://www.flowingdata.com for several), you can quickly learn how to use this software to prep data and create some stunning maps and visualization graphics. These graphics can then be exported for Web mapping. This tool has quickly become the go-to tool for Web mappers attempting to organize large amounts of data.

Other Data-Conversion and Data-Processing Tools

Hundreds, if not thousands, of free or affordable data-conversion and data-processing tools can be found online. Some do all of the work for you right in the cloud; others need to be downloaded and run on your computer. I do not necessarily endorse any software or technology over others. I only mentioned the ones with which I am most familiar. The book's accompanying Web site (http://www.ian.muehlenhaus.com/webcartography/) will provide links to additional resources as I am made aware of them.

A Web Cartographer's Tool Box

In this section, I provide a list of useful applications and (mostly) free APIs for designing interactive Web maps. Some of these are ideal for creating slippy maps, others for nonslippy maps, and yet others for stand-alone mobile map applications. This list is not exhaustive. However, it does represent some of the core technologies and APIs being put to use at the time of publication.

HTML5 and SVGs

As the use of mapping service APIs becomes easier and more common, it is easy to overlook one of the most effective and straightforward ways to represent thematic data on the Web: via good old Web graphics. One can create a simple Web page, embed an SVG map of countries or states, and style it based on its accompanying attribute data using CSS. One can add functionality and interactivity to the map (e.g., mouse over events, info windows) via simple JavaScript coding. Not only does designing maps this way provide you with absolute control over how your map looks—including what features are included and the projection—but it requires absolutely no interaction with external APIs. Many maps created by professional news organizations are designed in this fashion. Stand-alone SVG—embedded in a Web page without an API—often results in a more minimalist-looking and stylized map. Best of all, you can design these Web maps for absolutely free.

JavaScript and SVG maps are excellent in the following scenarios:

- For thematic representations;
- For designing maps with unconventional, noncylindrical/rectangular projections;

- For controlling map graphics that you created yourself in a design application (e.g., a cartogram designed in Adobe Illustrator exported as an SVG); and
- For embedding your maps within a Web page in more nuanced and complicated ways than might be possible with API-based maps.

Avenza MAPublisher

Perhaps one of the most overlooked mapping tools available today is Avenza MAPublisher (http://www.avenza.com). Avenza MAPublisher is a plug-in but not for Web browsers. Instead, it is a program that runs within Adobe Illustrator. It adds many GIS and cartography tools to the already-powerful graphic design capabilities of the Adobe Illustrator program. With Avenza MAPublisher, you can use Adobe Illustrator as a GIS, data organizer, data styler, and Web mapmaker without needing to learn any coding whatsoever. Avenza MAPublisher allows you to export any maps you create in Illustrator as interactive HTML5 Web maps and GeoPDFs. The HTML5 Web maps are slippy maps. You choose which layers to make selectable and decide which attribute information you would like included in the info windows. It does all of the programming for you (Figure 12.10). You just paste the code into your favorite HTML editing software and upload. Voilà!

FIGURE 12.10

Avenza MAPublisher is a plug-in that runs within Adobe Illustrator. The term *plug-in* does not do it justice, though. It essentially turns Adobe Illustrator into a GIS and Web mapmaking machine. You can import all sorts of data and design entire print and Web maps right in Illustrator. Then, you can export your print map with a slippy interface and info windows into either Adobe Flash (SWF) or HTML5 format to be uploaded to the Web. I created many of the Germany maps in Chapter 9 with Avenza MAPublisher. Here, it is about to be exported as an HTML5 interactive map.

Another neat feature that Avenza MAPublisher offers is the ability to export GeoPDFs (Figure 12.11). Avenza now has a mobile app, PDF Maps, for Apple iOS and Google Android devices that allows map users load GeoPDFs (see Figure 12.6). Moreover, Avenza has set up a separate marketplace within the app so you can sell your maps to mobile users. You can simply export your Avenza MAPublisher map from Adobe Illustrator as a GeoPDF and upload your map to their market. You can distribute your map to the world for free or charge for it; the choice is yours. The key feature of GeoPDFs, again, is that they do not need to be connected to the Internet to work. The map, interactive and slippy, is downloaded in its entirety to someone's device. Then, if a map user needs your map while sailing off the coast of Antarctica, it still works without an Internet connection and remains GPS enabled.

Avenza MAPublisher software may be of particular interest to you if you

- Want to design Web maps that are individually styled and have a more classic, print appearance than what many APIs offer;
- Want to design slippy maps that you have total control over just as you would when designing the layout and style of a paper map;
- Want maps that preserve your original projections and do not default to a Web Mercator projection;

FIGURE 12.11
Avenza MAPublisher also allows you to export your Adobe Illustrator maps as GeoPDFs. Again, these are PDFs (portable document files) that have geospatial data. Thus, they work with devices that have GPS. (See Figure 12.6 for an example of what the end product looks like on an iPad.)

- Want to design GeoPDF maps using all of the design features that Adobe Illustrator has; and
- Want to quickly supplement a print map you designed with an accompanying Web map or GeoPDF version for your customers.

Adobe Flash and Adobe AIR

Although the Web plug-in is slowly withering, I remain a staunch fan of Adobe Flash Professional (http://www.adobe.com). Adobe Flash Professional is not a plug-in. It is an interactive graphic design tool, and there are many things that Adobe Flash Professional allows a cartographer to do more easily and better than many other technologies. This is particularly true when Flash is used to create stand-alone applications (i.e., not for the Adobe Flash Player plug-in) that run using Adobe AIR.

Adobe AIR (which stands for Adobe Integrated Runtime) is a cross-platform system that runs programs made using Adobe Flash, Apache Flex, HTML, and Ajax. It allows you to create a program once and run it across all sorts of devices, including Windows and Mac computers, and numerous mobile operating systems. Flash Professional integrates with Adobe AIR so that you can immediately export your map in a variety of formats that run on all of these devices. Using ActionScript (the program language that Flash Professional and Adobe AIR use), you can create dynamic maps that load external datasets in real time just as you can with JavaScript and HTML5.

The shortcomings confronting the use of Adobe Flash Professional are abundant. As previously mentioned, Adobe has discontinued the Flash Player plug-in for most mobile Web browsers. Moreover, many map service companies have discontinued their Adobe Flash and ActionScript-based APIs. Although newer iterations of Adobe Flash Professional allow one to export animations as HTML5, the functionality of this feature is limited. Finally, Adobe Flash Professional and Player are not open source.

Shortcomings aside, Adobe Flash Professional is an excellent tool if you want to make Web maps in the following situations:

- Take static maps that exist as Adobe Illustrator files and add interactivity—even pan and zoom features;
- Create multimedia-rich, self-contained map apps to distribute or sell on mobile app markets (see Muehlenhaus, 2011, for a tutorial on making stand-alone Google Android apps with this software);
- Create maps the old-fashioned way, by drawing them (albeit on a digital art board); and
- Create dynamic and very robust map animations quickly and easily without needing to code.

Oracle Java

Often confused with JavaScript, Oracle Java has no relation to the former (http://www.java.com). In fact, JavaScript borrowed the Java name when Oracle Java was the dominant interactive Web development technology. In my opinion, Oracle Java is going the way of Adobe Flash when it comes to Web implementation. Around the time of this writing, the Oracle Java plug-in has been hampered by numerous security issues and is in a state of frequent patching. Concurrently, because Oracle Java works as a plug-in within the browser, and HTML5 is nearing complete implementation, APIs that use Java on the Web are increasingly being deprecated.

That being said, Oracle Java is one of the most widely known and used programming languages in the world. It can be used to design stand-alone map applications that run on different operating systems, including mobile apps. The Oracle Java plug-in may start to wither in browser distribution, but its utility for designing mobile map apps remains high. The Oracle Java programming language is useful for

- Creating mobile map apps for distribution on mobile markets and
- Creating map applications that run within PC browsers using the plug-in.

ESRI ArcGIS Online and API

ESRI has produced a very user-friendly and intuitive online mapping experience that can be broken down into two parts: the ArcGIS Online cloud mapping application and the ArcGIS JavaScript API.

ArcGIS Online (http://www.arcgis.com) is a wonderful mapmaking application. (Note: It is not entirely free, however, as you must pay money to import datasets with more than a thousand features.) For those of you who use GIS and have much of your data stored as shapefiles, you can upload them here without the need to convert them to another format first. (The site accepts other data formats as well.) Within ArcGIS Online, you can design your maps, change base maps, and conduct a variety of other GIS-like functions. Finally, you can embed your final map in your Web page, share it with others via a Web link, or later manipulate your Web map in a Web page using the ArcGIS JavaScript API.

The ArcGIS JavaScript API (http://help.arcgis.com/en/webapi/javascript/arcgis/) is the second additional component of ESRI's online mapping environment. Although it operates just like most other APIs, I find its online documentation and tutorials more user friendly than many of the open-source ones. This API is quite intuitive and ideal for people just starting to get their feet wet with Web mapping technologies. ESRI has even implemented a map element and GUI wizard of sorts that allows you to see how your map interfaces will look before you write the code.

These services will prove particularly useful in the following situations if

- You are an ESRI product user and have most of your data in shape-file format;
- You want to create a mobile Web map and have it automatically realign itself and resize map elements to fit a variety of screen sizes and resolutions;
- You do not want to design map elements from scratch but simply wish to use CSS to modify the style of premade map elements;
- You do not plan to use datasets with more than a thousand features each. (Again, you can pay to import larger datasets.)

TileMill Application and MapBox API

TileMill (http://www.mapbox.com/tilemill) is free open-source software that you download to your desktop to make slippy maps (see Figure 12.12). TileMill makes it fairly easy to create interactive maps using a modified version of CSS called CartoCSS. You can import standard data types, layer your data, and style the layers with nothing but CartoCSS. You can even make the styles change dynamically based on zoom level and other user input. Only one layer in your dataset can be made interactive, with mouse over and click functionality. There is a simple legend design tool as well.

Although not as full featured as many other map design tools, this is a great piece of software with which to become acquainted with CSS. Plus, you do not need to know any programming languages when you start to make your interactive Web maps work. TileMill exports maps in a variety of formats. The most popular export format is as MapBox Tiles.

MapBox (http://www.mapbox.com) is an online API and tile server tool. It can be used in conjunction with TileMill to host your TileMill maps on the Web. However, it also has its own API that can be interacted with using JavaScript to design completely interactive maps on one's own Web site. The goal of MapBox and its TileMill program is to help bring Web mapping to the masses. Many professional and government organizations have turned to TileMill to create professional-looking slippy maps.

TileMill is a great tool for quickly creating slippy maps in the following circumstances:

- When you are simply making a point-of-interest map showing different places on the map with info windows providing more information; and
- When you want to create customized and heavily stylized tile maps from scratch.

FIGURE 12.12

This is an image from the TileMill application. TileMill is a very powerful free Web map application created by MapBox (http://www.mapbox.com) that you can run on your desktop. It uses CartoCSS to style and design data that you have imported. You simply add shapefiles or other types of data and style it in the right-hand column. It is a very fast way to make interactive slippy maps. (Copyright MapBox. Map:© OpenStreetMap contributors, available for free under the Open Database License. Reproduced with permission.)

CartoDB Web Site and API

CartoDB is a spatial database management system that allows you to make styled and animated maps from your data tables. The Web site (http://www.cartoDB.com) offers you the ability to upload five different spatial datasets for free. One of the major advantages of using this service and its accompanying API is that you can easily create stellar info windows and uniquely styled data that can then be used in another map application designed with a different toolset. Also, the thematic design tools are top rate; you can quickly represent your spatial data thematically. Figure 12.13 shows proportional symbols created within seconds of loading a Natural Earth dataset (http://www.naturalearthdata.com). It does all of the work for you. Using CartoCSS, you can style all components of your map in some detail. This Web mapping service is extremely robust and will be of particular use to those who are

- Designing thematic Web maps;
- Interested in using more stylized info windows than are typically common in Web map APIs but not interested in doing all of the programming necessary to stylize them; and
- Adept at using CartoCSS to style map elements, as CartoDB makes use of this language as well.

Google Maps API

There is not much I can write about Google Maps here that has not been written about more thoroughly elsewhere. Google Maps truly revolutionized online mapping. Google offers, arguably, the most detailed publicly available dataset in the world. Their JavaScript API is very advanced and allows the manipulation of myriad map components. Moreover, as a technology services juggernaut, Google offers a variety of productivity APIs (https://code.google.com/apis/console) that can work together nearly seamlessly on a single Web site. In my mind, it is well worth learning JavaScript simply so you can play with the Google Maps API.

Separate from, but increasingly tied to, Google Maps, are Google Fusion tables. Google Fusion tables are currently a beta feature of Google Drive (an online map productivity suite). You can think of Google Fusion tables as the Formula One cars of spreadsheets. They are cloud based and are increasingly being used for real-time data storage and mapping.

Google Fusion tables allow you to import spreadsheets (e.g., CSV [comma-separated values] files) and then map, graph, and work with the data via the Google Maps API. I do not dare write too much more about the software here, as it is still in a beta version, and the interface seems to change frequently. If you already have a Google account, I highly recommend you explore the mapping and data-organizing potential of Google Fusion. I provide several links to tutorials dealing with Fusion tables at the end of this chapter.

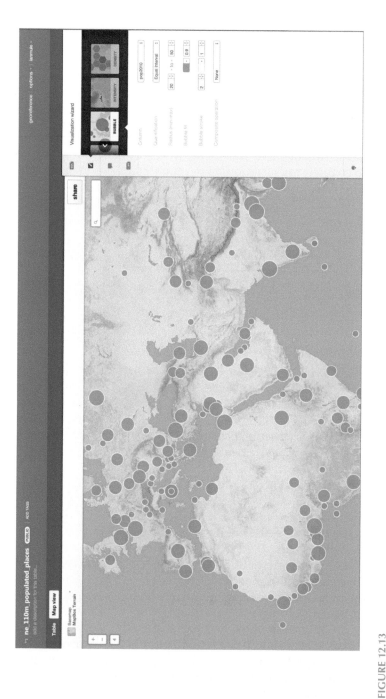

FIGURE 12.13

If you want to make stunning thematic representations in a hurry, I highly recommend CartoDB. CartoDB is a robust cloud mapping service. It facilitates CartoCSS, but it also has many built-in visualization schemes. Finally, in my opinion it has the best info window design options at the time of publication. I highly recommend you visit this site. (©CartoDB. Used with permission.)

The Google Maps API is great for the following Web mapping projects:

- Slippy maps for Web browsers;
- Stand-alone mobile map apps;
- Specialized reference map design; and
- Web maps that need location-based services and features.

CloudMade Leaflet API

CloudMade (http://www.cloudmade.com) is a company offering a variety of Web mapping tools, many of which are user friendly and when not free are extremely affordable. One of their most useful tools, and quickly becoming one of the new standards in Web mapping, is the open-source Leaflet API (see Figure 12.14). This API is designed to work with OpenStreetMap—although the API works with many other mapping services as well. What makes this API unique is that it is specifically designed with mobile mapping in mind, meaning that there are interactive controls designed specifically for post-WIMP (windows, icons, menus, pointer) interfaces. Leaflet automatically reformats layouts and map elements when screen real estate becomes constrained. The main benefit of using Leaflet is that you are using an open-source API to tweak open-source map tiles. The Leaflet API is, in my estimation, likely to become a worthy competitor to many of the proprietary map APIs flourishing right now.

The Leaflet API is an excellent choice for almost all of your slippy map-making needs, including

- Mobile map app development;
- Standard PC Web browser maps; and
- Mobile Web browser maps.

Other APIs

There are several challenges when reviewing different Web mapping technologies in a book. First, one cannot possibly know about all of them or pay each adequate attention. Second, APIs are constantly changing. Many of the technologies I have written about in the preceding pages have gone through numerous changes since I started writing. I am even loath to include the Web addresses of all of these tools and programs, as they may change. A simple Web search will find any of these in seconds.

Thus, although I only cover the APIs I am familiar with in these pages, I will attempt to update this book's Web site (http://www.ian.muehlenhaus. com/webcartography) with additional information about ones that are missing or new ones that arise in the future. For example, I can tell you right now that Microsoft Bing Maps and MapQuest have stellar APIs. I simply have not used them myself. So, please do not limit your exploration of data types and Web mapmaking tools only to the formats and applications mentioned in this chapter. Explore the Web; you are sure to find new tools to help you design your maps just the way you want.

Closing Advice

Everything I have written about in this chapter is now partially out of date. By now, there is probably a new Web mapping file type that is all the rage and being adopted by everyone who is anyone in Web mapping circles to create choropleth representations. Or, perhaps there is now a technology that allows you to simply draw your interactive map elements using a pen tool, hit a magic button, and add any interactivity you want. Here is my advice: Do not worry about it. The main point of Web mapping is not to use the latest and greatest technologies. The point is to create maps that communicate clearly and

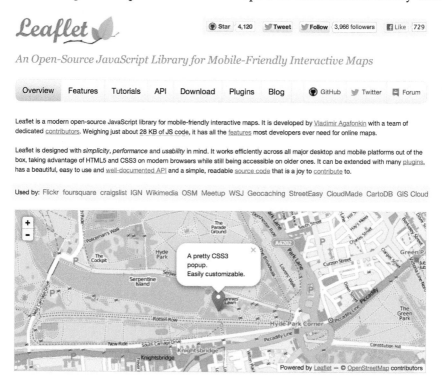

FIGURE 12.14
Leaflet JS by CloudMade is fast becoming the API of choice for many Web cartographers. It is open, free, and very robust. It also has good documentation. (©CloudMade. Used with permission.)

intuitively and that people can readily access from myriad interfaces. That is all that matters. Everything else is merely bragging fodder for Twitter and Facebook.

Cartography and mapmaking are forever evolving, just as all forms of communication are. The primary goals of communication remain the same regardless of medium and regardless of technology; they are to inform, persuade, and explain. It does not matter how you do it. A Web map made using Flash is of no less inherent value than one made using HTML5. All that should matter to a mapmaker is how well the communication is designed and how effectively your map presents the information to your intended audience. Anyone who critiques maps based on which technologies were used to create them is missing the whole point of mapmaking.

If there is one takeaway to leave you with it is this: *You do not design maps to use technology; you use technology to design better maps.*

Key Concepts

- You need to learn HTML, CSS, and JavaScript to truly harness the power of Web mapping.
- APIs are useful tools for incorporating myriad Web services into a single Web map.
- SVGs are a great way to design thematic maps without a Web Mercator projection or an API.
- There are numerous online map services that not only allow you to create Web maps without much coding but also, just as importantly, allow you to import desktop GIS data (e.g., shapefiles) and export the data in a Web map-friendly format (e.g., GeoJSON, KML).
- Explore new Web technology to create better maps as the technologies are constantly evolving.

Further Reading and Resources

Online Resources and Tutorials

Web Cartography Companion Web Site (http://www. ian.muehlenhaus.com/webcartography)

I have set up a companion Web site for this book (http://www.ian.muehlenhaus. com/webcartography). This Web site will provide links to inspirational Web maps, technical Web mapping tutorials (created by both myself and others), new developments in Web mapping, and a variety of other content. There is also special content for those of you who purchased this book.

Flowing Data (http://www.flowingdata.com)

The FlowingData Web site is great for free tutorials on a variety of random visualization practices. Of particular interest might be the tutorial on Chernoff faces and making maps with the free R program.

Lynda.com (http://www.lynda.com)

Lynda provides an incredible array of online video courses on nearly every type of Web technology, design, database, and productivity software and even on things such as time and project management. (The Lynda Time Management course with David Crenshaw pretty much saved my life while writing this book.) You can subscribe to the site by the month and gain unlimited access to all of their tutorials. You can peruse the tutorials they have available for free by visiting the site. I highly recommend this site for learning scripting and new Web technologies in a very short period of time. In my opinion, it is well worth the subscription fee.

W3Schools (http://www.w3schools.com)

This *free* Web site is a cornucopia of free tutorials, demos, and data. You can learn all of the basics about nearly every type of Web development and coding on this site. There is even a tutorial on using the Google Maps JavaScript 3 API. It is a great place to go to figure out the basics of HTML, CSS, XML, and JavaScript. I cannot recommend this site enough. My students use it religiously.

Recommended Tutorial Guides and Books

The Missing Manual Series by O'Reilly

O'Reilly's Missing Manual Series is a selection of tutorial and reference books on all types of software and hardware. They are written in layperson English and contain highly effective learning exercises that reinforce what you have read. You can sit down and read a book straight through (for example, I own one on CSS that I read this way) or you can use these books as references in the future when you forget something (I use my JavaScript book this way). They have a book on HTML 5, CSS, JavaScript, Flash, and many other topics.

There are many other great tutorial books. Coupled with YouTube or Lynda tutorials, you would be amazed how quickly you will be able to figure out just enough coding to make awesome Web maps.

From Print to Mobile mApps Tutorial

Muehlenhaus, I. (2012). From print to mobile mApps: How to take Adobe Illustrator maps, add pinch-to-zoom, and place them on the Android market. *Cartographic Perspectives 69*, 59–70.

I wrote a tutorial demonstrating how easy it is to take a static Adobe Illustrator map and turn it into an Android app to sell on Google Play (then the Android

Market). It is a step-by-step demonstration, with directions on setting up a Google Merchant account as well. Fortunately, it was published in a great journal, *Cartographic Perspectives*, that puts the copyright in the Creative Commons. So, you can access and reproduce this for free: http://cartographicperspectives.org/index.php/journal/article/view/cp69-muehlenhaus-oth/html (short URL: http://goo.gl/aAraD).

Visualize This by Nathan Yau

Yau, N. (2011). *Visualize this: the FlowingData guide to design, visualization, and statistics*. New York: Wiley. This book only has one chapter on mapping, but it is a worthwhile addition to every Web cartographer's library. This book is a tour de force of tutorials on how to use different pieces of software to make information graphics—both interactive and not. It covers everything from Python scripting, to using R to organize and visualize data, to Adobe Illustrator, to SVGs, to Adobe Flash, and to JavaScript and even Chernoff faces. It is very light on theory, but that is the only critique I can think of. It is a great introduction to some of these programs.

Wisconsin State Cartographer Web Site (http://www.sco.wisc.edu)

The Wisconsin State cartographer's Web site is an excellent resource for a variety of mapping needs. Of most pertinence here is its Learning Center (http://www.sco.wisc.edu/mapping-topics/learning-center.html). Here, you will find extremely useful tutorials and short white papers concerning a variety of Web mapping topics. If you happen to be looking for historical air photography or spatial data from Wisconsin, this is a great resource as well. The University of Wisconsin–Madison, where the State Cartographer's Office is located, has been and remains at the forefront of Web cartography. Keep your eye on this site as the information and tutorials are updated often.

Further Reading

Cartographic Perspectives Journal (http://www.cartographicperspectives.org)

Cartographics Perspectives Journal is the flagship journal of the North American Cartographic Information Society (NACIS; http://www.nacis.org). The journal recently went online only, and most of the content is free to everyone. What makes this academic journal special is that it caters to a wide audience, particularly professional map designers with little to no academic experience in cartography. Most important for your purposes, it typically has several technical and tutorial pieces in each issue devoted to Web mapping techniques and tricks. It is well worth checking out this journal, including past issues, on a regular basis.

Cartography and Geographic Information Science Journal (CaGIS)

The *Cartography and Geographic Information Science Journal* is not free, but you can subscribe if you join the organization by the same name (http://www.cartogis.org). (The membership fee is quite reasonable, and it is worth it for the journal.)

This journal is one of only a few official journals of the International Cartographic Association (http://www.icaci.org). Although the topics covered in the journal extend well beyond Web mapping, it is in this journal that many of the most prominent academics studying Web mapping publish.

Online Maps with APIs and Web Services, Edited by Michael P. Peterson

Peterson, M. P. (Ed.). (2012). *Online maps with APIs and Web services.* Berlin: Springer-Verlag. One book I highly recommend for those just getting started with Web services and APIs is the edited volume by Michael Peterson: *Online Maps with APIs and Web Services.* In this book, many of the leading academic experts on Web mapping discuss a variety of topics. Surprisingly, for a book of such academic content, almost all of the chapters are written in layperson, nontechnical English. Some of the topics covered include the latest trends in vector mapping on the Web; best practices for avoiding symbol clustering on your maps; using Adobe Flex (i.e., ActionScript) with the ArcGIS Flex API; and a great overview of using the Google Maps JavaScript 3 API while loading external data.

Index